ENVIRONMENTAL REMEDIATION TECHNOLOGIES,
REGULATIONS AND SAFETY

REMOVAL OF TOXIC CR(VI) FROM WASTEWATER

ENVIRONMENTAL REMEDIATION TECHNOLOGIES, REGULATIONS AND SAFETY

Additional books in this series can be found on Nova's website under the Series tab.

Additional e-books in this series can be found on Nova's website under the e-book tab.

ENVIRONMENTAL SCIENCE, ENGINEERING AND TECHNOLOGY

Additional books in this series can be found on Nova's website under the Series tab.

Additional e-books in this series can be found on Nova's website under the e-book tab.

ENVIRONMENTAL REMEDIATION TECHNOLOGIES,
REGULATIONS AND SAFETY

REMOVAL OF TOXIC CR(VI) FROM WASTEWATER

TONNI AGUSTIONO KURNIAWAN

Nova Science Publishers, Inc.
New York

For permission to use material from this book please contact us:
Telephone 631-231-7269; Fax 631-231-8175
Web Site: http://www.novapublishers.com

NOTICE TO THE READER

The Publisher has taken reasonable care in the preparation of this book, but makes no expressed or implied warranty of any kind and assumes no responsibility for any errors or omissions. No liability is assumed for incidental or consequential damages in connection with or arising out of information contained in this book. The Publisher shall not be liable for any special, consequential, or exemplary damages resulting, in whole or in part, from the readers' use of, or reliance upon, this material. Any parts of this book based on government reports are so indicated and copyright is claimed for those parts to the extent applicable to compilations of such works.

Independent verification should be sought for any data, advice or recommendations contained in this book. In addition, no responsibility is assumed by the publisher for any injury and/or damage to persons or property arising from any methods, products, instructions, ideas or otherwise contained in this publication.

This publication is designed to provide accurate and authoritative information with regard to the subject matter covered herein. It is sold with the clear understanding that the Publisher is not engaged in rendering legal or any other professional services. If legal or any other expert assistance is required, the services of a competent person should be sought. FROM A DECLARATION OF PARTICIPANTS JOINTLY ADOPTED BY A COMMITTEE OF THE AMERICAN BAR ASSOCIATION AND A COMMITTEE OF PUBLISHERS.

Additional color graphics may be available in the e-book version of this book.

LIBRARY OF CONGRESS CATALOGING-IN-PUBLICATION DATA

Removal of toxic Cr (VI) from wastewater / editor, Tonni Agustiono Kurniawan.
p. cm.
Includes bibliographical references and index.
ISBN 978-1-62081-025-5 (hardcover)
1 Sewage--Purification--Chromium removal. 2. Chromium-plating--Waste disposal. 3. Electroplating industry--Thailand--Bangkok--Environmental aspects. I. Kurniawan, Tonni Agustiono. II. Title: Removal of toxic chromium VI from wastewater. III. Title: Removal of toxic hexavalent chromium from wastewater.
 TD758.5.H43R468 2011   628.3'58--dc23

2012002166

Published by Nova Science Publishers, Inc. †New York

CONTENTS

Foreword		vii
Preface		ix
Acknowledgments		xi
List of Abbreviations		xiii
List of Units		xv
Chapter 1	Introduction	1
Chapter 2	Literature Review	11
Chapter 3	Analytical Methodology	81
Chapter 4	Results and Discussion	95
Chapter 5	Conclusion and Recommendations	147
References		155
Appendices		177
About the Author		183
Index		185

FOREWORD

The presence of toxic Cr(VI) in the aquatic environment and their harmful effects on living organisms has emerged as one of the most serious environmental concerns in Thailand, located in the Southeast Asian region. The research, undertaken at Sirindhorn International Institute of Technology of Thammasat University (2001-2003), directly addressed the need of our society for 'clean water'. Today one sixth of the population or almost 1.2 billion people living in developing countries are still without access to clean drinking water, while about 2.6 billions of the populations do not have access to basic sanitation facilities. About 400 millions are children under five-year old and elderly groups. This situation stands to worsen as our water supply has been contaminated by a wide variety of pollutants from industrial sources.

This study, which reports the technical applicability of coconut shell waste, a low cost material from local agricultural waste for treatment of Cr(VI)-contaminated water, can be a useful resource for environmental scientists, engineers, researchers, policy and decision makers, and engineering students worldwide. They will find insightful and novel ideas from the book how the solid waste, which presents disposal problems to the environment, could be chemically modified in such a way that it could be further used as a low cost material adsorbent to remove toxic Cr(VI) from contaminated wastewater, thus improving environmental protection. The study presented in this book was internationally recognized as the best thesis submitted to The BioInfo Bank Institute (Poland) in 2008.

PREFACE

In this study, the technical feasibility of various low cost adsorbents such as coconut shell charcoal (CSC) and zeolite as well as commercial activated carbon (CAC) for Cr(VI) removal from contaminated wastewater was investigated using batch studies at an initial chromium concentration from 5 to 10 mg/L and/or column operations.

Surface modifications of CSC and CAC using chitosan and oxidizing agents such as sulfuric and/or nitric acids respectively were also conducted to improve their removal performance, while pretreatment of zeolite with NaCl was performed. Using real wastewater, their removal performances on Cr are evaluated and statistically compared. Both the Langmuir and the Freundlich model of isotherms were employed to understand the adsorption mechanisms and kinetics of Cr removal by all the adsorbents.

It is evident that the chemical modifications of these adsorbents have significantly improved their removal capabilities for target metal. Both the CSC and CAC, which have been oxidized with nitric acid, have comparable Cr adsorption capacities (CSC: 3.22 mg/g, CAC: 6.04 mg/g) to those of synthetic wastewater at 10 mg/L of Cr concentration (CSC: 3.74 mg/g, CAC: 6.66 mg/g, zeolite: 3.55 mg/g). It is important to note that the Cr adsorption capacities of all types of the adsorbents vary from one to another, depending on the physico-chemical characteristics of the individual adsorbent, the extent of surface modifications, and the initial concentration of chromium.

ACKNOWLEDGMENTS

This thesis is based on research work undertaken at the Environmental Technology Program at Sirindhorn International Institute of Technology (SIIT) in Thammasat University (Thailand) under the supervision of Associate Prof. Dr. Sandya Babel. The author would like to express his gratitude to her for inspiring the author with stimulating ideas and useful suggestions throughout his study period (2001-2003).

Special thanks are also credited to Dr. Viboon Sricharoenchaikul and Dr. Alice Sharp as co-supervisors for providing him with valuable suggestions on this thesis. The author also would like to express his profound gratitude to Prof. Dr. Kensuke Fukushi (University of Tokyo, Japan) not only for serving as his thesis examiner, but also for his inspiring feedbacks and useful suggestions in improving the quality of his thesis.

Last but not least, the author's deepest gratitude is directed toward the Japanese Government, who awarded him the ADB-JSP scholarship, enabling him to pursue and complete his study at SIIT. His profound gratitude is also expressed to the German Federal Ministry for Education and Research (BMBF) for its fellowship during his research stay at the Ravensburg-Weingarten University of Applied Sciences (Germany).

Finally, the author dedicates this work not only to The Lord Jesus Christ for His abundant blessings, but also to his beloved mother and wife, whose unconditional love and persistent prayer has continuously supported him over the past years of study. What are the best in him, he owes to their credit alone.

LIST OF ABBREVIATIONS

ANOVA	Analysis of variance
APHA	American Public Health Association
BV	Bed volume
CAC	Commercial activated carbon
CSCCC	Coconut shell charcoal coated chitosan
COD	Chemical oxygen demand
Cr	Chromium
CSBAC	Coconut shell-based activated carbon
CSC	Coconut shell charcoal
GAC	Granular activated carbon
HSD	Honestly significant difference
RE	Regeneration efficiency
REA	Removal efficiency of adsorbent
SPSS	Statistical package for social science
The U.S.A	The United States of America

LIST OF UNITS

g	Gram
h	Hour(s)
kg	Kilogram
μg	Microgram
l	Liter
meq	Milliequivalent
mg	Milligram
mg/g	Milligram per gram
mg/l	Milligram per liter
ml	Milliliter
mm	Millimeter
ppm	Part per million
rpm	Rotation per minute
v/v	Volume/volume

Chapter 1

INTRODUCTION

ABSTRACT

This chapter introduces the problems of water pollution which have occurred in Bangkok (Thailand) partly due to the generation of Cr(VI)-contaminated water from local electroplating industries. It is divided into four sections. The first section describes the problems of a number of electroplating industries that generate a large amount of Cr(VI)-contaminated wastewater in recent years, and their potential impact on public health and the environment. The second section provides an overview of various environmental technologies that have been employed worldwide in recent decades for treatment of heavy metals-contaminated water and elaborates their advantages and limitations in applications. This section also identifies the existing research gaps in treatment of Cr(VI)-contaminated water that will be bridged through the implementation of this project. The third section presents several specific objectives that will be achieved through laboratory studies in batch and column modes to meet the increasing strict requirements of effluent limit imposed by local environmental legislation. The fourth section elaborates the scope of the present study and its approaches to accomplish the objectives outlined in this study.

In general, this chapter aims to establish and justify the need for this study to bridge the research gaps in the area of water pollution control. In this regard, adsorption treatments using commercial activated carbon (CAC) and various low cost adsorbents such as coconut shell charcoal (CSC) and zeolite will be tested. The project is expected to be able to improve the treatability of Cr(VI)-contaminated water from local electroplating industries and their treated effluents may be able to comply with the increasingly strict requirements imposed by local environmental legislation.

1.1. PROBLEM IDENTIFICATION

Electroplating industry has been playing significant roles in Thailand since the early part of the 20th century. It has a unique position as an industry characterized by its massive employment. The electroplating industry has demonstrated an outstanding growth due to the development of large numbers of metal manufacturers and engineering industries throughout the country.

Presently, there are more than 2,000 units registered electroplating industries in Thailand, out of which about 95% units are situated in Bangkok. There are also large numbers of unregistered electroplating units located outside Bangkok, but their number is still unknown. The local electroplating industries in Thailand have proliferated from small plating shops serving small manufacturers, such as jeweler makers to sub-unit of complex enterprises, such as automobile industry manufacturers and metal finishing. Since these plating shops could have managed to meet the demands of their customers and obtained a sizeable profit, they have survived in large numbers until today.

Currently, there are more than 95% units electroplating industry in Bangkok adopting a conventional type of chrome-plating process for electroplating (Kongsricharoern, 1994). Consequently, a large volume of Cr-contaminated water is generated by the metal plating industry, where a typical daily production of a small enterprise (with less than 100 employees) is about 1,000 m^3. For this reason, in recent years the generation of chromium-contaminated water has become one of the most critical problems arising from the local electroplating industries in Thailand.

Chromium, which is on the top priority list of Group A toxic pollutants defined by the U.S. EPA (Environmental Protection Agency) based on its chronic effects, is present in the electroplating wastewater as Cr(VI) in the form of oxyanions, such as chromates (CrO_4^{2-}), dichromates ($Cr_2O_7^{2-}$), and bichromates ($HCrO_4^-$) (Kotaś and Stasicka, 2000). Depending on the pH and total Cr(VI) concentration (Tandon et al., 1984; Ramos et al., 1994; Lee et al., 1995), the Cr(VI) species exist with equilibrium constants (at 25°C) as follows:

$$H_2CrO_4 \leftrightarrow H^+ + HCrO_4^- \qquad k_1 = 0.18 \qquad (pK_1 = 6.51) \qquad (1.1)$$

$$HCrO_4^- \leftrightarrow H^+ + CrO_4^{2-} \qquad k_2 = 3.2 \times 10^{-7} \qquad (pK_2 = 5.65) \qquad (1.2)$$

$$2HCrO_4^- \leftrightarrow Cr_2O_7^{2-} + H_2O \quad k_3 = 33.3 \qquad (pK_3 = 14.56) \qquad (1.3)$$

The public concern about this heavy metal stem from the fact that Cr(VI) is highly toxic to living organisms compared to Cr(III) ions (Selvaraj et al., 2003). Due to its high solubility in the aquatic environment, the Cr(VI) can be readily adsorbed by living organism. Once Cr(VI) enters the food chain, a large concentration of Cr compounds will accumulate in animals and human body. If total chromium is ingested beyond the permitted maximum concentration (0.1 mg/L), it can cause various types of acute and chronic health disorders, such as nausea, diarrhea, vomiting, hemorrhaging, epigastric pain, and chromosomic aberrations (Lalvani et al., 1998; Gupta et al., 1999). Therefore, it is necessary to completely remove Cr(VI) from industrial wastewater before being discharged into natural water streams or onto land.

According to Kongsricharoern (1994), the Cr(VI) concentration in the wastewater effluents discharged from local electroplating industries in Bangkok was about 25 mg/L. So far, their current treatment facilities using chemical precipitation method still produce Cr effluent of 10 mg/L, far higher than the acceptable limit of the wastewater discharge standard allowed by Thai government and the U.S.EPA, which are 0.25 mg/L (Thai Pollution Control Department, 2003) and 0.05 mg/L (Nourbakhsh et al., 1994; Selvaraj et al., 2003) respectively. Undoubtedly this toxic metal has become one of the most important factors causing water pollution in Chao Phraya River in Bangkok. Therefore, the presence of plating shops has threatened not only the surrounding environment, but also the life of people living in Bangkok due to water pollution caused by regular plating waste production day after day continuously.

The metal plating shops in Bangkok are now facing an urgent problem due to the increasingly strict environmental regulations imposed by the Thai government. They are looking for a proper treatment method for their Cr-contaminated wastewater before being discharged into the environment. A cost-effective environmental technology for the treatment of Cr-contaminated water is highly desired by industrial users, especially those living in developing countries like Thailand.

1.2. RATIONAL OF STUDY

In order to neutralize the toxic effects of Cr(VI) on living organisms and the environment, recently various techniques have been employed to treat metal-contaminated water. These methods include electrochemical precipitation (Kongsricharoen and Polprasert, 1995, 1996), flocculation (Li et al., 2003), ion exchange (Dobrevski et al., 1996; Sapari et al., 1996; Rengaraj et al., 2001; Lin and Kiang, 2003), ultrafiltration/ nanofiltration (Ahn et al., 1999; Vrijenhoek and Waypa, 2000; Brandhuber and Amy, 2001; Tavares et al., 2002; Yurlova et al., 2002), and reverse osmosis (Slater et al., 1983; Benito and Ruiz, 2002; Qin et al., 2002; Ning, 2002). All of these treatment methods have their advantages and limitations in applications.

Due to its simple equipments, convenient operation, and capability of treating a large volume of wastewater containing high concentration of heavy metal, traditionally chemical reduction and precipitation is the most commonly employed method in Thailand for Cr(VI) removal in the past decades (Charerntanyarak, 1999). Unlike other heavy metals, since Cr(VI) cannot be simply precipitated as insoluble hydroxides, its reduction to the trivalent state should be followed by alkaline precipitation of Cr(III) hydroxide at pH 9.0-10.0. In spite of its widespread usage in local electroplating industry for Cr treatment, major drawbacks with chemical precipitation include an excessive sludge production that requires further treatment, slow metal precipitation kinetics that requires the provision of detention times of over 30 minutes for metal precipitation, poor settling characteristics of metal precipitates, aggregation of metal precipitates, the increasing cost of solid waste disposal for the sludge generated, and most importantly, the long-term environmental consequences for the local environment where the sludge is disposed (Ouki et al., 1997a; Bose et al., 2002). Therefore, chemical precipitation is not the most preferable method for treating Cr-contaminated water generated from local electroplating industries.

Ion exchange is considered a better alternative technique for Cr(VI) treatment than chemical precipitation, as it does not present sludge disposal problems (Dobrevski et al., 1996). This technique is widely used for treatment of metal-contaminated water in industries in developed countries, such as Japan, the U.S., and the Europe Union. Ion exchange is a proven technology, as resins are available with highly selective metal ion, easy to use, and less time-consuming. Furthermore, ion exchange offers potentials of valuable metal recovery that can reduce the concentration of target metal ions to a very low level. However, its limitation for treating inorganic effluent is that appropriate pretreatment systems for secondary effluents are required. The removal of suspended solids from wastewater should be done before conducting ion exchange. Last, but not least, other drawbacks are that ion exchange is not selective enough for chromate ions over foreign anions and that it needs high capital investment and operational costs. This situation makes this particular treatment less attractive to electroplating industries in

Bangkok due to its high operational cost. Consequently, there is a growing need among local industrial users to develop other viable treatment methods, which can effectively remove Cr from electroplating wastewater at lower treatment cost.

Any proposed treatment method should use low-cost chemicals, abundantly available and local unused resources, simple design of installation and maintenance. The treatment method should be also sludge-free operation, environmentally friendly, safe, economically viable, technically feasible, easily applicable, highly efficient for Cr removal with potentials of metal recovery, easy in terms of handling and controlling the operations, and most importantly, capable of producing treated effluents that can meet the increasingly strict requirements imposed by local environmental regulations.

In order to accomplish all these requirements, in recent years adsorption treatment method has been developed to be the most effective and feasible option for wastewater treatment. Basically, adsorption is a mass transfer process where a substance is transferred from the liquid phase to the surface of a solid and become bound by physical and/or chemical process. Due to its physical properties, large surface area, microporous structure, high adsorption capacity, and high degree of surface reactivity, adsorption using granular commercial activated carbon (CAC) has been gaining a considerable attention from environmental technologists for treating electroplating wastewater containing toxic heavy metals, such as Cd (Ramos et al., 1997), Ni (Shim et al., 2001), Cr (Ouki et al., 1997a), and Cu (Monser and Adhoum, 2002) with metal adsorption capacities of 8, 10, 32, and 38 mg/g of CAC, respectively.

By using the adsorption treatment method, it may be possible to meet the stringent standard of Cr(VI) effluent discharge (less than 0.25 mg/L) in Thailand. However, in spite of the versatility of CAC for wastewater treatment application and its prolific use for the removal of heavy metals in the past decades (Ik and Zoltek, 1977; Huang and Wu, 1977; Huang, 1977; Huang and Bowers, 1978; Netzer and Hughes, 1984; Reed and Arunachalam, 1994a, 1994b; Reed et al., 1995; Candela et al., 1995; Seron et al., 1996; Sharma and Forster, 1996a; Ramos et al., 1994, 2002; Aggarwal et al., 1999; Han et al., 2000; Chen and Wang, 2000; Leinonen and Lehto, 2001; Park and Jung, 2001; Babić et al., 2002; Faur-Brasquet et al., 2002; Hu et al., 2003), CAC remains an expensive material, considering that the better the quality of CAC, the higher its cost.

To replace costly CAC, the search for low cost adsorbents has intensified in recent years. Natural materials that are available in large quantities or certain waste products from agricultural operations may have potentials as low cost adsorbents, as they represent unused resources. For this reason, in recent years various non-conventional materials have been investigated to achieve an economically feasible and effective treatment of Cr(VI)-contaminated water. It was reported that chitosan (Udaybhaskar et al., 1990; Schmuhl et al., 2001), sphagnum moss peat (Sharma and Forster, 1993, 1994a, 1995), leaf mould (Sharma and Forster, 1996b), peat (Dean and Tobin, 1999), bark (Alves et al., 1993), used tyres (Hamadi et al., 2001), kyanite (Ajmal et al., 2001), agricultural waste (Ajmal et al., 2000), coconut shell (Alaerts et al., 1989), tannin (Nakano et al., 2001), sawdust (Selvi et al., 2001), rice husk (Srinivasan et al., 1988), soya cake (Daneshvar et al., 2002), moss (Lee et al., 1995), banana pith (Low et al., 1995), diatomite (Dantas et al., 2001), oxide (Gupta and Tiwari, 1995; Bailey et al., 1992; Weng et al., 1997), red mud (Pradhan et al., 1999a), fly ash (Panday et al., 1984a, 1984b, 1985), iron(III) hydroxide (Namasivayam and Ranganathan, 1993; Aoki and Munemori, 1982), fly ash (Bayat, 2002), coal (Lakatos et al., 2002), wool (Balkaya, 2002), distillery sludge (Selvaraj et al., 2003), and sugar beet pulp (Reddad et al., 2003) can bind Cr.

Due to its capability of removing undesirable heavy metals at low cost, recently coconut shell charcoal (CSC) has been developed into one of the newer promising options for the removal of heavy metals. Coconut shell (*Cocos nucifera Linn*), which in many cases often presents serious problems of disposal for local environments, is an abundantly available agricultural waste from local coconut industries. According to the statistics compiled by the Thai Ministry of Commerce (1998), about 1.4 million tonnes of coconut (*Cocos nucifera*) are produced in Thailand every year. Since coconut shell has little economic value, it can be obtained for free or at minimal cost. Conversion of coconut shell, representing an unused resource, into activated carbon, which can be used as an adsorbent in water purification, would add its economic value, help reduce the cost of waste disposal, and most importantly, provide a potentially inexpensive alternative to costly CAC.

Coconut shell, generated in the separation process of fibers from a nutshell, is a hard material. Morphologically, it encloses a thick bony endocarp (shell). Coconut shell, which can be further used for the production of granular CSC, is made up of 35% cellulose, 28% hemicellulose, 25% lignin, and the remainder being other constituents. Its exchange properties are attributed to the presence of surface functional groups, such as carboxylic, hydroxyl carbonyl, phenolic, oxyl, hydroxyl, and lactone, which have a high affinity for heavy metal ions, such as cadmium, chromium, and cobalt (Tan et al., 1993). Metal uptake is believed to occur through sorption process involving the surface functional groups mentioned previously. Treatment method for Cr-contaminated water using CSC appears to be technically feasible, environmentally friendly, and economically attractive, as its current commercial price is one-forth of the price of CAC.

Preliminary studies have been conducted to asses the technical feasibility of using coconut shell-based activated carbon to remove certain metal ions, such as Cr(VI) (Alaerts et al., 1989), Cd(II) and Pb(II) (Arulanantham et al., 1989) from contaminated water. However, these previous studies did not provide any information on the surface modifications of CSC to improve its removal performance on heavy metal. In the present study, CSC was chemically modified with chitosan and/or a strong oxidizing agent, such as nitric acid and/or sulfuric acid, respectively. Using real wastewater, the Cr removal performance of CSC was evaluated and statistically compared to that of CAC.

Zeolite, a crystalline aluminosilicates from natural resources, is also an inexpensive material, as its commercial price is a half of the price of CAC. Due to its high cation exchange capacities (CEC) with certain metal ions in the solution, recently zeolite has received a considerable interest for heavy metal removal (Zamzow et al., 1990). The exchange properties of zeolite are due to the presence of its negatively charged lattice, which is exchangeable with certain metals in the solution. Since the applications of zeolite on industrial scale was limited to the removal of cesium and strontium from radioactive waste (Semmens and Martin, 1988), the use of zeolite for Cr removal from contaminated water was explored in this research.

1.3. OBJECTIVE OF STUDY

In this laboratory studies, batch and column operations were conducted to investigate the technical feasibility of using CAC and low cost adsorbents such as CSC and zeolite for Cr(VI) removal from contaminated water. The specific objectives are presented as follows:

1) To find out the optimum conditions required to achieve a maximum removal of all adsorbents for chromium in batch studies;
2) To assess and compare the adsorption capacity of all types of adsorbents in batch studies;

3) To investigate the ability of treated effluents to meet the increasingly strict Cr limits imposed by local environmental legislation and the US EPA;

4) To study the regeneration for all types of spent CSC and/or CAC in column experiments using a combined regenerant of NaOH subsequently followed by HNO_3; and

5) To study the adsorption mechanisms of Cr(VI) by CSC and/or zeolite in the aqueous solutions.

1.4. Scope of Study

Investigations on the technical feasibility of all adsorbents for Cr(VI) removal were divided into two phases: batch and column studies. Batch experiments were conducted using real wastewater. The optimum conditions of pertinent factors such as: dose, pH, agitation speed, and contact time, were determined from batch studies.

In order to evaluate the practical implications of the results of Cr removal performance obtained from different types of adsorbents and to obtain the operational parameters for industrial application, batch studies were also performed using real wastewater originated from a local electroplating industry in Rangsit (Thailand). Both the Langmuir and Freundlich isotherms were used to understand the adsorption mechanism and kinetics of Cr(VI) removal from the aqueous solution. Cost effectiveness of using CSC and/or zeolite as an adsorbent compared to CAC for Cr(VI) removal are also discussed in the present study, assuming that operational costs, such as chemicals, electricity, labor, transportation, and maintenance cost, are the same for all types of adsorbents.

Using real electroplating wastewater, column studies were also conducted to validate the results of batch experiments and evaluate the technical feasibility of all types of adsorbents for the possibility of industrial applications. In order to return the adsorbent into its original condition for subsequent uses and to restore its adsorption capacity without losing the adsorbent's initial structure, column regeneration was also conducted using various chemicals. A combination of sodium hydroxide directly followed by nitric acid was used as regenerants for spent CSC and/or CAC, while sodium hydroxide alone was used to regenerate spent zeolite.

Surface modifications of CSC and/or CAC with chitosan and/or strong oxidizing agents, such as sulfuric and/or nitric acids were also performed to improve their removal performances on Cr. Chemical treatment of zeolite with sodium chloride was conducted to prepare the adsorbent in the homoionic form of Na^+ prior to ion exchange with Cr^{3+} in the solution at acidic conditions. The results of the Cr removal performances of all types of adsorbents in as-received and chemically modified forms were evaluated and statistically compared using paired t-test and/or analysis of variance (ANOVA) as well as Tukey's multiple comparison tests at the confidence interval of 95% ($p \leq 0.05$).

It is important to note that all the experiments were carried out in duplicate in order to ensure the accuracy, reliability, and reproducibility of the collected data and the mean values of two data sets are presented. The criteria assigned for the relative standard deviation of the Cr removal efficiency for all experiments was less than 2.0%. When the relative error exceeded this criterion, the data were disregarded and a third experiment would be conducted until the relative error fell within an acceptable range.

Chapter 2

LITERATURE REVIEW

ABSTRACT

This chapter reviews the most up-to-date research on various environmental technologies for metals decontamination. It also discusses and assesses the technical feasibility of a variety of physico-chemical treatments including chemical precipitation, flotation, membrane filtration, coagulation-flocculation, adsorption and electrochemical technique that have been used in recent years for removal of heavy metals from aquatic environment. The results of their removal performance on different types of heavy metals are evaluated and compared to each other.

This chapter is divided into a number of sections. The first section describes local water pollution problems caused by heavy metals. The second section presents various physico-chemical treatments for treatment of heavy metals-contaminated wastewater and critically assesses their treatment performance for target metals. The third section identifies prominent physico-chemical techniques and compares their treatment costs.

This chapter concludes with a summary of outstanding techniques for treatment of heavy metals-contaminated water. The author's literature survey of 193 reported studies (1977-2003) within the body of knowledge demonstrates research gaps in the area of environmental science and technology that needs to be bridged through the implementation of this project. In this present study, coconut shell charcoal (CSC) and zeolite are employed to address the existing research gaps in the area of water treatment. The results of the present study reported in Chapter 4 are likely to improve the treatability of Cr(VI)-contaminated wastewater from local electroplating industries in Bangkok (Thailand). Their treated effluents may be able to comply with increasingly strict requirements imposed by the Thai environmental legislation.

2.1. HEAVY METALS-CONTAMINATED WASTEWATER

Due to the discharge of large amount of metal-contaminated wastewater, electroplating industry is one of the most hazardous among the chemical-intensive industries (Babel and Kurniawan, 2004). Inorganic effluent from the industries contains toxic metals such as Cd, Cr, Cu, Ni and Zn (Kurniawan, 2003), which tend to accumulate in the food chain.

Because of their high solubility in the aquatic environments, heavy metals can be absorbed by living organisms. Once they enter the food chain, large concentrations of heavy metals may accumulate in the human body. If the metals are ingested beyond the permitted concentration, they can cause serious health disorders (Table 2.1). Therefore, it is necessary to treat metal-contaminated wastewater prior to its discharge to the environment.

Table 2.1. MCL of heavy metals in surface water and their toxicities

Heavy Metal	Toxicities
Cr(VI)	Headache, nausea, diarrhea, vomiting
Cr(III)	
Zn(II)	Depression, lethargy, neurologic signs such as seizures and ataxia
Cu(II)	Liver damage, Wilson disease, insomnia
Cd(II)	Kidney damage, renal disorder, Itai-Itai
Ni(II)	Dermatitis, nausea, coughing

Different treatment techniques for wastewater laden with heavy metals have been developed in recent years both to decrease the amount of wastewater produced and to improve the quality of the treated effluent. Although various treatments such as chemical precipitation, coagulation-flocculation, flotation, ion exchange and membrane filtration can be employed to remove heavy metals from contaminated wastewater, they have their inherent advantages and limitations in application.

Chemical precipitation is widely used for the treatment of electroplating wastewater in Thailand (Charerntanyarak, 1999) and Turkey (Tünay and Kabdasli, 1994). Coagulation-flocculation has also been employed for heavy metal removal from inorganic effluent in Thailand (Charerntanyarak, 1999) and China (Li et al., 2003). Sorptive flotation has attracted interest in Greece (Lazaridis et al., 2001) and the US (Doyle and Liu, 2003) for the removal of non-surface active metal ions from contaminated wastewater. In recent years, ion exchange has also received considerable interest in Italy (Pansini et al., 1991) and Spain (Ayuso and Sánchez, 2003) as one of the most promising methods to treat wastewater laden with heavy metals.

Due to its convenient operation, membrane separation has been increasingly used recently for the treatment of inorganic effluent. There are different types of membrane filtration such as ultrafiltration (UF), nanofiltration (NF) and reverse osmosis (RO). Membrane filtration has been used in Taiwan (Juang and Shiau, 2000) and Korea (Ahn et al., 1999) to remove Cd(II), Ni(II), Zn(II) and Cr(III) ions from contaminated wastewater.

Electrotreatments such as electrodialysis (Pedersen, 2003), membrane electrolysis (Janssen and Koene, 2002) and electrochemical precipitation (Kongsricharoern and Polprasert, 1995) have also contributed to environmental protection. However, these techniques have been investigated less extensively due to the high operational cost caused by energy consumption. Although many techniques can be employed for the treatment of inorganic effluent, each treatment has its inherent limitation. The ideal treatment should be not only suitable, appropriate and applicable to the local conditions, but also able to meet the maximum contaminant level (MCL) standards established (Table 2.1) (Babel and Kurniawan, 2003a).

This chapter presents an overview of the technical applicability of a variety of physico-chemical techniques such as chemical precipitation, coagulation-flocculation, membrane filtration, flotation and adsorption for treating wastewater laden with heavy metals. Their advantages and limitations in application are evaluated. To highlight their technical applicability for this purpose, selected information on pH, dose, initial metal concentration, and treatment efficiency is discussed. Low-cost adsorbents that stand out for high metal adsorption capacities are presented and compared to activated carbon.

2.2. PHYSICO-CHEMICAL TREATMENT TECHNIQUES FOR WASTEWATER LADEN WITH HEAVY METALS

2.2.1. Chemical Precipitation

Chemical precipitation is widely used for heavy metal removal from inorganic effluent (Benefield and Morgan, 1999; US EPA, 2000). After pH adjustment to the basic conditions (pH 11), the dissolved metal ions are converted to the insoluble solid phase via a chemical reaction with a precipitant agent such as lime (Wang et al., 2004). Typically, the metal precipitated from the solution is in the form of hydroxide (Tünay, 2003). The conceptual mechanism of heavy metal removal by chemical precipitation is presented in Equation (2.1) (Wang et al., 2004):

$$M^{2+} + 2\,(OH)^- \leftrightarrow M(OH)_2 \downarrow \qquad (2.1)$$

where M^{2+} and OH^- represent the dissolved metal ions and the precipitant, respectively, while $M(OH)_2$ is the insoluble metal hydroxide.

Lime precipitation was employed for the removal of heavy metals such as Zn(II), Cd(II) and Mn(II) cations with initial metal concentrations of 450, 150 and 1,085 mg/L, respectively, in a batch continuous system (Charerntanyarak, 1999). In spite of their varying initial concentrations, an almost complete removal from synthetic wastewater was achieved for all the metals at pH 11, complying with the effluent limit of the Thai Pollution Control Department for Zn(II) and Mn(II) of less than 5 mg/L (PCD, 2005). However, the treated effluent was unable to meet the stringent limit set by the US EPA of lower than 1 mg/L (US EPA, 2005), thus suggesting that subsequent treatments using other physico-chemical methods were still required to comply with the US EPA discharge standard.

Some attractive findings were reported by Tünay and Kabdasli (1994), who studied the applicability of hydroxide precipitation in a closed system to treat synthetic wastewater containing Cd(II) and Cu(II) ions. Inorganic cations (Ca(II) and Na) were employed as ligand-sharing agents for EDTA (ethylenediamine-tetraacetic acid) and NTA (nitrilotriacetic acid). They reported that Ca(II) was the only cation that effectively bound both ligands to form the hydroxide precipitations of the complexed metals. At pH 11, EDTA was also found to be the major component that determined Cd(II) solubility (Tünay et al., 1994).

Different results were obtained for the removal of Ni(II) uptake from a low-strength of real wastewater with a metal concentration of less than 100 mg/L (Papadopoulos et al, 2004). At pH 7.5 and 10.5, the researchers found that about 71% and 85% of Ni(II) removal, respectively, with an initial metal concentration of 51.6 mg/L, could be attained. This could be attributed to the fact that a greater portion of the Ni(II) was precipitated and removed in the form of insoluble hydroxide compounds with the increasing pH.

Overall, pH adjustment to the basic conditions (pH 11) is the major parameter that significantly improves heavy metal removal by chemical precipitation. Due to its availability in most countries, lime or calcium hydroxide is the most commonly employed precipitant agent. Lime precipitation can be employed to effectively treat inorganic effluent with a metal concentration of higher than 1,000 mg/L. Other advantages of using lime precipitation include the simplicity of the process, inexpensive equipment requirement, and convenient and safe operations, making it a popular method for metal removal from contaminated wastewater.

In spite of its advantages, chemical precipitation requires a large amount of chemicals to reduce metals to an acceptable level for discharge (Jüttner et al., 2000). Other drawbacks are its excessive sludge production that requires further treatment, the increasing cost of sludge disposal, slow metal precipitation, poor settling, the aggregation of metal precipitates, and the long-term environmental impacts of sludge disposal (Yang et al., 2001; Bose et al., 2002; Wingenfelder et al., 2005).

2.2.2. Coagulation-Flocculation

Coagulation-flocculation can be employed to treat wastewater laden with heavy metals. Principally, the coagulation process destabilizes colloidal particles by adding a coagulant and results in sedimentation (Shamas, 2004). To increase the particle size, coagulation is followed by the flocculation of the unstable particles into bulky floccules (Semerjian and Ayoub, 2003). The general approach for this technique includes pH adjustment and involves the addition of ferric/alum salts as the coagulant to overcome the repulsive forces between particles (Licskó, 1997).

After lime precipitation, Charerntanyarak (1999) employed subsequent coagulation process to remove Zn(II), Cd(II) and Mn(II) ions from synthetic wastewater. The optimum pH for coagulation process was found to be 11. At pH 11, the concentration of Zn(II) and Mn(II) in the treated effluent was reduced to less than 5 mg/L, the limit for the wastewater discharge set by the Thai Pollution Control Department (2005).

To treat real electroplating wastewater containing copper, Li et al. (2003) modified the conventional coagulation-flocculation process using sodium diethyldithiocarbamate (DDTC) as a trapping agent and both poly-ferric sulphate and polyacrylamide as the flocculants. DDTC is the most common chemical used as metal precipitant to form insoluble metaldithio salts (Andrus, 2000). These insoluble dithiometal salts are then coprecipitated, forming hydroxide-neutralized solids that precipitate prior to the discharge of the treated waste stream. When the mole ratio of DDTC to Cu was between 0.8

and 1.2, they found that an almost complete removal of Cu(II) could be achieved.

In general, coagulation-flocculation can treat inorganic effluent with a metal concentration of less than 100 mg/L or higher than 1,000 mg/L. Like chemical precipitation, pH ranging from 11.0 to 11.5 has been found to be effective to improve the heavy metal removal by the coagulation-flocculation process. Improved sludge settling, dewatering characteristics, bacterial inactivation capability, sludge stability are reported to be the major advantages of lime-based coagulation (Edwards, 1994; Cheng et al., 1994).

In spite of its advantages, coagulation-flocculation has limitations such as high operational cost due to chemical consumption. The increased volume of sludge generated from coagulation-flocculation may hinder its adoption as a global strategy for wastewater treatment. This can be attributed to the fact that the toxic sludge must be converted into a stabilized product to prevent heavy metals from leaking into the environment (Ayoub et al., 2001).

To overcome such problems, electro-coagulation may be a better alternative than the conventional coagulation, as it can remove the smallest colloidal particles and produce just a small amount of sludge (Vik et al., 1984; Elimelech and O'Melia, 1990). However, this technique also creates a floc of metallic hydroxides, which requires further purification (Persin and Rumeau, 1989), making the recovery of valuable heavy metals impossible.

2.2.3. Flotation

Flotation is employed to separate solids or dispersed liquids from a liquid phase using bubble attachment (Wang et al., 2004). The attached particles are separated from the suspension of heavy metal by the bubble rise. Flotation can be classified as: (i) dispersed-air flotation, (ii) dissolved-air flotation (DAF), (iii) vacuum air flotation, (iv) electroflotation and (v) biological flotation. Among the various types of flotation, DAF is the most commonly used for the treatment of metal-contaminated wastewater (Zabel, 1984). Adsorptive bubble separation employs foaming to separate the metal impurities. The target floated substances are separated from bulk water in a foaming phase.

Laboratory study was carried out by Rubio and Tessele (997) to investigate the flotation of Zn(II) and Ni(II) from synthetic wastewater using chabazite as the adsorptive particulate. They found that the removal performance was dependent on the interfacial chemistry and aggregation effectiveness. An almost complete removal (98.6%) of heavy metal ions with an initial concentration of 2 mg/L could be achieved by using 20 mg/L of $Fe(OH)_3$. The results are comparable to those of Blöcher et al. (2003), who combined flotation and membrane separations to remove Ni(II) cations from synthetic plating solution by using CTABr (cetyl trimethyl-ammonium bromide) as the cationic collector.

Other interesting results are reported by Zamboulis et al. (2004), who studied the sorptive flotation for the removal of Zn(II) and Cu(II) ions from synthetic wastewater. SDS (sodium dodecyl sulfate) and HDTMA (hexadecyl-trimethyl-amonium-bromide) were used as cationic collectors. The addition of 2 g/L of zeolite was found to remove 99% of 50 mg/L of zinc(II), while 4 g/L of zeolite was required to remove 97% of 500 mg/L of the initial Cu(II) concentration. The results confirmed that both the surface charge of the system and the solution pH significantly affected the metal removal by zeolite (Lazaridis et al., 2001).

To explore their application as biosurfactants, Surfactin-105 and Lychenysin-A were applied to enhance the removal of Cr(VI) and Zn(II) ions from synthetic wastewater (Zouboulis et al., 2003). An almost complete removal of both metals with initial metal concentrations of 50 mg/L could be achieved at pH 4.0.

In the past decades, the trends of research had shifted from flotation alone to a combination of flotation and other physico-chemical treatment such as filtration or powder activated carbon (Offinga, 1995; Lainé et al., 1998). Encouraging results were reported by Mavrov et al. (2003), who examined a newly integrated process that combined adsorption, membrane separation and flotation for Cu(II) and Zn(II) removal from real wastewater with synthetic zeolite as the bonding agent. They found that about 97% of Cu(II) and Zn(II) removal were attained with an initial metal concentration of 60 mg/L. The binding capacities of Zn(II), Cu(II) and Ni(II) ions were found to be 270 mg/g, 200 mg/g and 60 mg/g, respectively. The results were higher than those of Doyle and Liu (2003), who employed triethylenetetraamine (Trien) as the collector for the flotation of Cu(II) ions from synthetic wastewater.

Although it is only a kind of physical separation process, heavy metal removal by flotation has the potential for industrial application (Jokela and Keskitalo, 1999). Low cost materials such as zeolite and chabazite have been found to be effective collectors with removal efficiency of higher than 95% for an initial metal concentration ranging from 60-500 mg/L. Flotation can be employed to treat inorganic effluent with a metal concentration of less than 50 mg/L or higher than 450 mg/L. Other advantages such as a better removal of small particles, shorter hydraulic retention times and low cost make flotation one of the most promising alternatives for the treatment of metal-contaminated wastewater (Matis et al., 2003; 2004).

2.2.4. Membrane Filtration

Membrane filtration has received considerable attention for the treatment of inorganic effluent, since it is capable of removing not only suspended solid and organic compounds, but also inorganic contaminants such as heavy metals. Depending on the size of the particle that can be retained, various types of membrane filtration such as ultrafiltration, nanofiltration and reverse osmosis can be employed for heavy metal removal and are presented as follows:

2.2.4.1. Ultrafiltration(UF)

UF utilizes permeable membrane to separate heavy metals, macromolecules and suspended solids from inorganic solution on the basis of the pore size (5-20 nm) and molecular weight of the separating compounds (1,000-100,000 Dalton) (Vigneswaran et al, 2004). These unique specialties enable UF to allow the passage of water and low-molecular weight solutes, while retaining the macromolecules, which have a size larger than the pore size of the membrane (Sablani et al, 2001).

Some significant findings were reported by Juang and Shiau (2000), who studied the removal of Cu(II) and Zn(II) ions from synthetic wastewater using chitosan-enhanced membrane filtration. The amicon-generated cellulose YM10 was used as the ultrafilter. About 100% and 95% rejection were achieved at pH ranging from 8.5-9.5 for Cu(II) and Zn(II) ions, respectively, with an initial Cu(II) concentration of 79 mg/L and Zn(II) concentration of 81 mg/L. The results indicated that chitosan significantly enhanced metals removal by 6-10 times compared to using membrane alone. This could be attributed to the major role of the amino groups of chitosan chain, which served as coordination sites for metal binding. In acidic conditions, the amino groups of chitosan are protonated after reacting with H^+ ions as follows:

$$RNH_2 + H^+ \longleftrightarrow RNH_3^+ \qquad Log\ Kp = 6.3 \qquad (2.2)$$

Having the unshared electron pair of the nitrogen atom as the sole electron donor, the non-protonated chitosan binds with the unsaturated transition metal cation through the formation of coordination bond (Fei et al, 2005). For most of the chelating adsorbent, the functional groups with the donor atoms are normally attached to the metal ions, thus leading to a donor-acceptor interaction between chitosan and the metal ions (Navarro et al, 1998), as indicated by the Equation (2.3):

$$M^{2+} + n \, RNH_2 \leftrightarrow M-(RNH_2)n^{2+} \qquad (2.3)$$

where M and RNH_2 represent metal and the amino group of chitosan, respectively, while n is the number of the unprotonated chitosan bound to the metal. Combination of Equations (2.2) and (2.3) gives the overall reaction as follows:

$$M^{2+} + n \, RNH_3^+ \leftrightarrow M-(RNH_2)_n^{2+} + n \, H^+ \qquad (2.4)$$

Equation (2.4) suggests that an increase in pH would enhance the formation of metal-chitosan complexes.

To improve the rejection rates of metals by complexation-ultrafiltration, some modifications were conducted using polyethyleneimine (PEI), a water-soluble macroligand, to remove Cr(III) ions from a synthetic solution (Aliane et al., 2001). Technical parameters such as pH, ligand concentration, applied pressure and membrane pore size were found to significantly affect the rejection rate of metal ions. The researchers found that pH 6.0, pressure of 3 bar and 2 g/L of PEI were the optimum conditions to achieve the Cr rejection rate of 95% with an initial metal concentration of 20 mg/L.

Another significant breakthrough in UF research was explored by Yurlova et al (2002) for the removal of Ni(II) ions from a synthetic solution using micellar-enhanced UF. Both sodium dodecyl sulfate (SDS) and non-ionic monoalkylphenol polyetoxilate were used to form micelles. An almost complete removal of Ni(II) could be achieved at 4 bar of pressure with 1 g/L of SDS concentration. This result was higher than that of Akita et al. (1999), who studied the removal of Ni(II) ions from synthetic wastewater via micellar-enhanced UF. They found that only 60% of Ni(II) with an initial metal concentration of 29 mg/L was removed at pH ranging from 5 to 7, confirming that the metal rejection rates were dependent on the degree of complexation between the cations and the extractant within the micelle (Kryvoruchko et al., 2002).

To explore its potential to remove heavy metals, Saffaj et al. (2004) employed low cost $ZnAl_2O_4$-TiO_2 UF membranes for the removal of Cd(II) and Cr(III) ions from synthetic solution. They reported that 93% Cd(II) rejection and 86% Cr(III) rejection were achieved. Such high rejection rates might be attributed to the strong interactions between the divalent cations and the positive charge of the membranes. These results indicate that the charge capacity of the UF membrane, the charge valencies of the ions and the ion concentration in the effluent, played major roles in determining the ion rejection rates by the UF membranes (Laine et al., 2000).

Depending on the membrane characteristics, UF can achieve more than 90% of removal efficiency with a metal concentration ranging from 10-112 mg/L at pH ranging from 5-9.5 and at 2-5 bar of pressure. UF presents some advantages such as lower driving force and a smaller space requirement due to its high packing density. However, the decrease in UF performance due to membrane fouling has hindered it from a wider application in wastewater treatment. Fouling has many adverse effects on the membrane system such as flux decline, an increase in transmembrane pressure (TMP) and the biodegradation of the membrane materials (Choi et al., 2005). These effects result in high operational costs for the membrane system.

2.2.4.2. Nanofiltration(NF)

Nanofiltration has unique properties between UF and RO membranes. Its separation mechanism involves steric (sieving) and electrical (Donnan) effects. A Donnan potential is created between the charged anions in the NF membrane and the co-ions in the effluent to reject the latter (Bruggen and Vandecasteele, 2003). The significance of this membrane lies in its small pore and membrane surface charge, which allows charged solutes smaller than the membrane pores to be rejected along with the bigger neutral solutes and salts.

To evaluate polyvinyl alcohol as the skin materials of the NF membrane, Ahn et al. (1999) investigated the uptake of Ni(II) ions from real electroplating wastewater using NTR-7250 membranes. The researchers found that the removal of Ni(II) was dependent on the applied pressure and the initial metal concentrations. It was observed that beyond 2.9 bar of pressure, the removal of Ni(II) did not improve with the increasing pressure, suggesting that 2.9 bar was the optimum pressure for NF application to remove Ni(II) ions from wastewater.

A comparative study of the removal of Cu(II) and Cd(II) ions from synthetic wastewater using nanofiltration (NF) and/or reverse osmosis (RO) was conducted by Qdais and Moussa (2004). At the same initial metal concentration of 200 mg/L, 98% of Cu(II) removal and 99% of Cd(II) removal could be attained by using RO, while NF was capable of removing more than 90% of Cu(II) and 97% of Cd(II). These results indicate that both types of the membrane filtration are effective for metal removal from contaminated wastewater. However, NF requires a lower pressure than RO, making NF more preferable due to its lower treatment costs (Mohammad et al., 2004).

In general, NF membrane can treat inorganic effluent with a metal concentration of 2,000 mg/L. Depending on the membrane characteristics, NF can effectively remove metal at a wide pH range of 3-8 and at pressure of 3-4 bars. However, NF is less intensively investigated than UF and RO for the removal of heavy metal.

2.2.4.3. Reverse Osmosis (RO)

In reverse osmosis (RO), a pressure-driven membrane process, water can pass through the membrane, while the heavy metal is retained. This treatment has gained favor in Malaysia (Ujang and Anderson, 1996) and Spain (Benito and Ruiz, 2002). Due to the increasingly stringent environmental legislation, RO has been developed with a membrane pore size down to 10^{-4} μm (Bohdziewicz et al, 1999). By applying a greater hydrostatic pressure than the osmotic pressure of the feeding solution, cationic compounds can be separated from water (solvent).

To examine the quality of polyamide as the skin material of the RO membrane, the removal performance of an ultra-low-pressure reverse osmosis membrane (ULPROM) was investigated by Ozaki et al. (2002) for the separation of Cu(II) and Ni(II) ions from both synthetic and real plating wastewater. An almost complete rejection was attained for both Cu(II) and Ni(II) cations at 5 bar of pressure. It was pointed out that the higher the transmembrane pressure, the higher the flux and rejection rates. pH ranging from 7 to 9 was found to be optimum to achieve a maximum metal rejection.

Using polyamide as the same skin material for the membrane, Qin et al. (2002) also employed RO for the treatment of Ni-contaminated wastewater directly discharged from a metal plating industry. They found that 97% of Ni(II) removal was achieved. The rejection rate of Ni(II) was enhanced at pH ranging from 3.5 to 7.0. Such a phenomenon might be due to the Donnan exclusion mechanism of the charged membranes, as the membrane acquired significantly more negative charge at pH 7.0 than at pH 3.5, thus inducing high rejection rates through electrostatics attraction.

In general, compared to UF and NF, RO is more effective for heavy metal removal from inorganic solution, as indicated by the rejection percentage of over 97% with a metal concentration ranging from 21-200 mg/L. Depending on the characteristics of the membrane such as the porosity, material, hydrophilicity, thickness, roughness and charge of the membrane (Madaeni and Mansourpanah, 2003), RO works effectively at a wide pH range of 3-11 and at 4.5-15 bar of pressure. Unlike chemical precipitation, instead of pH, pressure is the major parameter that affects the extent of heavy metal removal by RO. The higher the pressure, the higher the metal removal efficiencies, and thus the higher the energy consumption. The other advantages of using RO include a high water flux rate, high salt rejection, resistance to biological attack, mechanical strength, chemical stability and the ability to withstand high temperatures (Ujang and Anderson, 1996). The reuse of water from industrial process can be achieved and RO enables industrial users to comply with the effluent limit of the discharge standards imposed under environmental legislation.

In spite of its benefits, RO has some limitations. Due to the suspended solids or oxidized compounds such as chlorine oxides, the small pores of the membrane make it more prone to fouling (Potts et al., 1981). Any cations such as Cd(II) and Cu(II) present in the contaminated wastewater promote membrane fouling, which might be irreversible. The membrane would then have to be replaced, thus increasing the operational costs. Membrane performance also decreases over time, resulting in the decreasing permeate flow rate (Ning, 2002). Other major drawbacks are the high energy consumption, the scaling of $CaCO_3$ or $CaSO_4$ and the need for experienced personnel to run the process (Slater et al, 1983).

2.2.5. Ion Exchange

In addition to membrane filtration, ion exchange is one of the most frequently applied treatments worldwide for wastewater laden with heavy metals. In ion exchange, a reversible interchange of ions between the solid and liquid phases occurs, where an insoluble substance (resin) removes ions from an electrolytic solution and releases other ions of like charge in a chemically equivalent amount without any structural change of the resin (Rengaraj et al, 2001; Vigneswaran et al, 2004). Ion exchange can also be used to recover valuable heavy metals from inorganic effluent (Dąbrowski et al, 2004). After separating the loaded resin, the metal is recovered in a more concentrated form by elution with suitable reagents.

Since the acidic functional groups of resin consist of sulfonic acid, it is assumed that the physicochemical interactions that may occur during metal removal can be expressed as follows:

$$nRSO_3^- — H^+ + M^+ \longleftrightarrow nRSO_3^- —M^{n+} + nH^+ \qquad (2.5)$$
(resin)   (solution)   (resin)   (solution)

where $(-RSO_3^-)$ and M represent the anionic group attached to the ion exchange resin and the metal cation, respectively, while n is the coefficient of the reaction component, depending on the oxidation state of metal ions (Dąbrowski et al, 2004).

The application of natural exchangers such as clinoptilolite and synthetic zeolite (NaPl) to purify synthetic wastewater was investigated by Álvarez-Ayuso et al (2003). It was found that synthetic zeolite demonstrated a sorption capacity about 10 times greater than that of clinoptilolite, despite having a comparable surface area (20-28 m^2/g). This could be attributed to the H^+ exchange capacity of zeolite and the strength of hydration shell cations that played major roles in the sorption capacities of both exchangers (Papadopoulos et al, 2004).

Other synthetic resins such as Amberlite IR-120 and Dowex 2-X4 were employed to investigate the total uptake of heavy metals from real plating wastewater containing Zn(II), Cr(III) and Cr(VI) ions (Sapari et al, 1996). The ion exchange system was found to completely remove all the heavy metals from the solution.

Similar results for Cr(III) uptake were also obtained by Rengaraj et al. (2001), who investigated the removal performance of IRN77 and SKN1 resins in a synthetic solution. They reported that a complete removal of Cr(III) with a higher metal concentration of 100 mg/L could be achieved. Other resins such as 1200H, 1500H and IRN97H were also employed to study the kinetics of Cr uptake from real and synthetic wastewater (Rengaraj et al, 2003).

Another investigation using Amberlite IR120 resin was performed on the uptake of multi-cations (Ni(II), Mn(II) and Co(II)) from complex synthetic solution containing EDTA, NTA and citrate (Ali and Bishtawi, 1997). Amberlite IR-120 is a strongly acidic resin with a sulfonic acid functionality. The researchers found that the equilibrium exchange of metal and resins was dependent on the pH and the complex agent used, confirming the results of other studies (Keane, 1998).

Synthetic resin such as Ambersep 132 was also explored by Lin and Kiang (2003) to recover chromic acid (H_2CrO_4) from synthetic plating solution. Both batch and column tests were carried out to compare the performance of the recovery process. In the batch studies, the Langmuir isotherm (92.1 mg/g) was more representative than the Freundlich (24.56 mg/g) for the equilibrium sorption data of chromic acid. The metal exchange capacity in the column operation (100 mg/g) was higher than that of the batch studies (92.1 mg/g) at the same concentration of 750 mg/L. This could be attributed to the fact that in the batch studies, the concentration gradient decreased with an increasing contact time; while in the column operation, the resin had continuous physicochemical contact with fresh feeding solution at the interface of the adsorption zone, when the adsorbate solution passed through the column. Consequently, more cations were exchanged in the column operation than in the batch studies (Ko et al, 2001; Kabay et al, 2003; Gode and Pehlivan, 2003; Dries et al, 2005).

In general, ion exchange is effective to treat inorganic effluent with a metal concentration of less than 10 mg/L, or in the range of 10-100 mg/L, or even higher than 100 mg/L. Depending on the characteristics of the ion exchanger, heavy metal removal by ion exchange works effectively in acidic conditions with pH ranging from 2-6. Among the ion-exchangers, low-cost material such as clinoptilolite can give a comparable metal removal (90%) to commercial resins such as IRN77 and SKN1 (100%) for the same Cr(III) concentration of 100 mg/L. Unlike chemical precipitation, ion exchange does not present any sludge disposal problems (Korngold et al, 2003), thus lowering the operational costs for the disposal of the residual metal sludge. Other advantages of ion exchange include its convenience for fieldwork since the required equipment is portable, the speciation results are reliable and the experiments can be done quickly. Resins also have certain ligands that can selectively bond with certain metal cations, making ion exchange easy to use and less time-consuming (Korngold et al, 2003).

Despite these advantages, ion exchange also has some limitations in treating wastewater laden with heavy metals. Prior to ion exchange, appropriate pretreatment systems for secondary effluent such as the removal of suspended solids from wastewater are required. In addition, suitable ion exchanger resins are not available for all heavy metals and the operational cost is high (Ahmed et al, 1998).

2.2.6. Electrochemical Techniques

2.2.6.1. Electrodialysis (ED)

Electrodialysis is a membrane separation in which ionized species in the solution are passed through an ion-exchange membrane by applying an electric potential (Bruggen and Vandecasteele, 2002). The membranes are thin sheets of plastic materials with either anionic or cationic characteristics. When a solution containing ionic species passes through the cell compartments, the anions migrate toward the anode and the cations toward the cathode, crossing the anion-exchange and cation-exchange membranes (Ito et al, 1980; Chen, 2004).

Some interesting results were reported by Tzanetakis et al. (2003), who evaluated the performance of the ion exchange membranes for the electrodialysis of Ni(II) and Co(II) ions from synthetic solution. Two cation exchange membranes, perfluorosulfonic Nafion 117 and sulfonated polyvinyldifluoride membrane (SPVDF), were compared under similar operating conditions. By using perfluorosulfonic Nafion 117, the removal efficiency of Co(II) and Ni(II) were 90% and 69%, with initial metal concentrations of 0.84 mg/L and 11.72 mg/L, respectively.

Different results were reported from a laboratory scale study of Cd(II) treatment from synthetic plating wastewater using ED (Marder et al, 2003). Two commercial cationic and anionic exchange membranes, Nafion 450 and Selemion, were employed. About 13% of Cd(II) with an initial metal concentration of 2 g/L was removed within 120 min.

The literature review above indicates that ED cannot effectively treat inorganic effluent with a metal concentration higher than 1,000 mg/L, thus suggesting that ED is more suitable for a metal concentration of less than 20 mg/L. In spite of its limitation, ED offers advantages for the treatment of wastewater laden with heavy metals such as the ability to produce a highly concentrated stream for recovery and the rejection of undesirable impurities from water. Moreover, valuable metals such as Cr and Cu can be recovered. Since ED is a membrane process, it requires clean feed, careful operation, periodic maintenance to prevent any damages to the stack.

2.2.6.2. Membrane Electrolysis (ME)

Membrane electrolysis, a chemical process driven by an electrolytic potential, can also be applied to remove metallic impurities from metal finishing wastewater. There are two types of cathodes used: a conventional metal cathode (electrowinning) and a high surface area cathode (Janssen and Koene, 2002). When the applied electrical potential across an ion exchange membrane, reduction-oxidation reaction takes place in electrodes (Chen et al, 2004). In the anode, oxidation reactions occur as follows:

$$M_1 \text{ (insoluble)} \leftrightarrow M_1^{n+} \text{ (soluble)} + n\,e^- \qquad (2.6)$$

$$4OH^- \leftrightarrow 2H_2O + O_2 + 4e^- \qquad (2.7)$$

$$2Cl^- \longleftrightarrow Cl_2 + 2e^- \qquad\qquad (2.8)$$

In the cathode, the following reduction reactions take place:

$$M_2^{n+} \text{ (soluble)} + n\ e^- \longleftrightarrow M_2 \text{ (insoluble)} \qquad (2.9)$$

$$2H^+ + 2e^- \longleftrightarrow H_2 \text{ (g)} \qquad\qquad (2.10)$$

M and n represent the metal and the coefficient of the reaction component, respectively. The n coefficient depends on the state oxidation of the metal ions.

The feasibility of electrochemical Cr(VI) removal from synthetic wastewater using carbon electrodes was investigated by Rana et al (2004). More than 98% of Cr removal with an initial metal concentration of 8 mg/L could be achieved at pH 2.0. This result is slightly lower than that of Martínez et al. (2004), who studied Cr(VI) removal from synthetic plating wastewater. An almost complete Cr removal could be achieved with an initial metal concentration of 130 mg/L, consuming 7.9 kWh/m^3 of energy. Martínez et al. (2004) also reported that the higher the current density, the shorter the treatment time required and the lower the energy spent for agitation.

Ni(II) recovery from synthetic rinse water of plating baths using ME was studied by Orlan et al (2002). About 90% of Ni(II) was recovered from an initial metal concentration of 2 g/L. However, this process required 4.2×10^3 kWh/m^3 of energy consumption, significantly higher than that reported by Kongsricharoern and Polprasert (1995; 1996), which consumed 12-20 kWh/m^3 using electrochemical precipitation.

Unlike ED, ME can be employed to treat plating wastewater with a metal concentration of higher than 2,000 mg/L or less than 10 mg/L. The major drawback of ME is its high energy consumption.

2.2.6.3. Electrochemical Precipitation (EP)

To maximize the removal of heavy metal from contaminated wastewater, electrical potential has been utilized to modify the conventional chemical precipitation. Some work using electrochemical precipitation (ECP) was carried out for the removal of Cr(VI) from real electroplating wastewater (Kongsricharoern and Polprasert, 1995). Over 80% of Cr removal could be attained and the treated Cr effluent was less than 0.5 mg/L, the effluent limit allowed in Thailand. The result is comparable to that of Kongsricharoern and Polprasert (1996), which also employed bipolar ECP for Cr(VI) removal using the same type of wastewater. Bipolar ECP was also technically applicable for 85% of Cr removal with an initial Cr concentration of 2,100 mg/L.

In general, electrochemical precipitation processes can treat inorganic effluent with a metal concentration higher than 2,000 mg/L. Depending on the characteristics of the electrodes, the electrochemical process can work at either acidic or in basic conditions (Subbaiah et al, 2002). Grebenyuk et al. (1989; 1996) reported that heavy metal removal can be carried out through electrochemical oxidation/reduction processes in an electrochemical cell without a continuous feeding of redox chemicals, thus avoiding a costly space, time and energy consumption.

2.2.7. Adsorption

Recently, adsorption becomes one of the alternative treatments (Kurniawan et al., 2005). Basically, adsorption is a mass transfer process by which a substance is transferred from the liquid phase to the surface of a solid, and becomes bound by physical and/or chemical interactions (Kurniawan and Babel, 2003). Due to its large surface area, high adsorption capacity and surface reactivity, adsorption using activated carbon can remove metals such as Ni(II) (Shim et al., 1996), Cr(VI) (Ouki et al., 1997), Cd(II) (Leyva-Ramos et al., 1997), Cu(II) (Monser and Adhoum, 2002) and Zn(II) (Leyva-Ramos et al., 2002), from inorganic effluent and the treated effluent can meet stringent standards for metal effluent discharge in some countries (Table 2.1). However, the use of activated carbon may be costly.

In recent years, the search for low-cost adsorbents that have metal-binding capacities has intensified. Materials locally available in large quantities such as natural materials, agricultural waste or industrial by-products can be utilized as low-cost adsorbents. Some of these materials can be used as adsorbents with little processing. Conversion of these materials into activated carbon, which can be used as an adsorbent for water purification, would improve economic value, helping industries reduce the cost of waste disposal and providing a potential alternative to activated carbon.

To achieve an economically effective treatment of metal-contaminated wastewater, various low-cost materials have been investigated worldwide, such as in India (Ajmal et al., 2001), Thailand (Kurniawan, 2002), Nigeria (Abia et al., 2003), Italy (Abollino et al., 2003) and USA (Yu et al., 2003). It was reported that wool (Balkaya, 2002), soya cake (Daneshvar et al., 2002), sawdust (Dakiky et al., 2002), maple saw dust (Yu et al., 2003), distillery sludge (Selvaraj et al., 2003), cocoa shell (Meunier et al., 2003), sugar beet pulp (Reddad et al., 2003) and zeolite (Babel and Kurniawan, 2004a) could bind Cr(VI) with high removal capacity.

In this particular section, a critical overview of various low-cost adsorbents derived from agricultural waste, industrial by-products or natural materials that have metal-binding capacity is presented.

2.2.7.1. Agricultural Waste

Agricultural waste is one of rich sources for low-cost adsorbents besides industrial by-product or natural material. Due to abundant availability, agricultural waste such as coconut shell and cocoa shell poses little economic value and moreover, creates serious disposal problems (Selvakumari et al., 2002). To achieve an economically effective treatment of metal-contaminated water, such unused resources can be used as an adsorbent for metal uptake.

Babel and Kurniawan (2004b) investigated the applicability of coconut shell charcoal (CSC) modified with oxidizing agents and/or chitosan for Cr(VI) removal using simulated wastewater. CSC is an agricultural waste from coconut industry in Thailand at US$0.25/kg. CSC oxidized with nitric acid had higher Cr adsorption capacities (10.88 mg/g) than that oxidized with sulfuric acid (4.05 mg/g) or coated with chitosan (3.65 mg/g). The results suggest that surface modification of CSC with a strong oxidizing agent generated more adsorption sites on its solid surface for metal adsorption. To make Cr removal by CSC more economical, further work has been done by Kurniawan (2002) to regenerate spent adsorbent. It was reported that desorption and regeneration of CSC with NaOH and HNO_3 still enabled the same column for multiple uses in subsequent cycle with regeneration efficiency of more than 95% (Kurniawan, 2002).

Cu(II) and Zn(II) removal from real wastewater were studied using pecan shells (Bansode et al., 2003). Some treated pecan shells used are: PSA (phosphoric acid-activated pecan shell carbon), PSC (carbon dioxide-activated pecan shell carbon); PSS (steam-activated pecan shell carbon). PSA and PSS had good removal capacities for both ions. The Freundlich isotherm was applicable for the equilibrium sorption of PSA, suggesting that metal uptakes ions took place on a heterogeneous surface by multilayer adsorption.

Orange peel was another agricultural waste used for Ni(II) removal from simulated wastewater (Ajmal et al., 2000). The maximum metal removal occurred at pH 6.0 and the adsorption followed the Langmuir isotherm, indicating that Ni(II) uptake might occur on a homogenous surface by monolayer adsorption. A metal adsorption capacity of 158 mg/g was achieved at 323K. This result is significantly higher than that of similar study undertaken by Annadurai et al. (2002), suggesting that the adsorption capacity of an adsorbent depends on the initial concentration of adsorbate.

A comparative study was conducted on Cu(II) removal from simulated solution using phosphoric-modified pecan shells and/or by a SR5 resin (Shawabkeh et al., 2002). At pH 3.6, the adsorption capacity of pecan shell on Cu(II) (95 mg/g) was higher than that of the resins (80 mg/g). At pH higher than 8.5, pecan shell had an adsorption capacity of 180 mg/g, almost two times higher than that at pH 3.6. This measured Cu(II) adsorption capacity was not a reliable result since, at pH higher than 8.5, Cu(II) ions precipitated in the form of hydroxide, thus increasing the metal removal from the solution.

Demirbaş et al. (2002) investigated Ni(II) removal from simulated solution using hazelnut shell. With an initial metal concentration of 15 mg/L, the optimum Ni(II) removal took place at pH 3.0 with metal adsorption capacity of 10.11 mg/g. In another study, hazelnut shell was also employed for Cr(VI) adsorption from simulated solution with an initial Cr(VI) concentration of 1,000 mg/L (Kobya, 2004). About 170 mg/g of Cr(VI) capacity occurred at pH 1.0. The results indicate that the adsorption capacity of individual adsorbent depends on the initial metal concentration.

Bishnoi et al. (2003) conducted a study on Cr(VI) removal by rice husk from aqueous solution. The maximum metal removal by rice husk took place at pH 2.0 and multilayer adsorption might occur on the surface of adsorbent, as indicated by the applicability of the Freundlich isotherm for the equilibrium data. However, Ajmal et al. (2003) reported that the Langmuir isotherm was representative for the metal equilibrium data of rice husk. These differences could be due to the extent of surface modification on the individual adsorbent. Bishnoi et al. (2003) employed formaldehyde for pretreatment of rice husk, while K_2HPO_4 was used in Ajmal et al. (2003). The two chemicals have different effects on the extent of surface modification that affects their reactivity in adsorbing heavy metal from solution.

The uptake of Cd(II), Ni(II) and Cu(II) ions from real industrial wastewater using coirpith was studied (Kadirvelu et al., 2001a). Maximum metal removal occurred at pH ranging from 4.0-5.0. Since coirpith was made up of homogenous adsorption patches, monolayer adsorption might occur on the surface, as indicated by the applicability of the Langmuir isotherm for the equilibrium sorption. This result is in agreement with the later study undertaken by Kadirvelu et al. (2001b), who used coirpith for Ni(II) uptake.

Hasar (2003) studied Ni(II) adsorption from simulated solution using almond husk. Maximum metal removal of 37.17 mg/g occurred at pH 5.0.

Monolayer adsorption might occur on the adsorbent surface, as indicated by the applicability of the Langmuir isotherm for the equilibrium sorption. Although both originated from almond waste, the Ni(II) adsorption capacity of almond husk (37.17 mg/g) was almost four times higher than that of almond shell (10 mg/g) (Dakiky et al., 2002). This can be due to the fact that the cell walls of almond husk contain higher concentration of cellulose, silica and lignin than those of almond shell. Consequently, almond husk has more hydroxyl groups and carboxylic groups than almond shell for metal adsorption, resulting in higher metal removal by almond husk (Hasar, 2003).

Rice hull, containing cellulose, lignin, carbohydrate and silica, was investigated for Cr(VI) removal from simulated solution (Tang et al., 2003). To enhance its metal removal, the adsorbent was modified with ethylenediamine. The maximum Cr(VI) adsorption of 23.4 mg/g was reported to take place at pH 2. This result indicates that an acidic pH range was effective for Cr(VI) removal from the solution. In acidic solution, Cr(VI) demonstrates a very high positive redox potential (E^o) in the range of 1.33-1.38 V (Kotaś and Stasicka, 2000), implying that Cr(VI) was unstable in the presence of electron donors. The carbon surface of rice hull contains carboxylic and hydroxyl groups, which play a role as electron donors in the solution (Tan et al., 1993). Consequently, Cr(VI) oxyanion is readily reduced to Cr(III) ions due to the presence of electron donors of rice hulls according to the following reaction at pH 2.0-6.0:

$$HCrO_4^- + 7\,H^+ + 3e^- \leftrightarrow Cr^{3+} + 4\,H_2O \quad E^o = 1.20\ V \qquad (2.11)$$

This can be observed that the reduction of Cr(VI) oxyanion is accompanied by proton consumption in the acidic solution, confirming the role played by H^+ in the Cr(VI) removal.

Thioglycolic acid-modified cassava waste was explored for the removal of Cd(II), Cu(II) and Zn(II) from simulated solution (Abia et al., 2003). After chemical modification, adsorption kinetics was rapid and equilibrium was attained within 20 min. The adsorption capacities of the adsorbent were 18.05, 11.06 and 56.82 mg/g for Cd(II), Zn(II) and Cu(II) ions, respectively; thus suggesting that the surface modification of cassava waste with thio-glycolic acid improved its metal performance.

Enhanced Cu(II) adsorption was achieved using as-received soybean hull (ASH) and citric acid-modified soybean hull (CMSH) (Marshall et al., 1999). Chemical modification of soybean hull with citric acid remarkably improved its metal removal (ASH: 24.76 mg/g; CMSH: 154.9 mg/g). This could be due to the fact that pretreatment increased not only the number of carboxyl groups, but also the negative charges on the hulls.

Peanut hulls were investigated for Ni(II) removal from synthetic solution (Periasamy and Namasivayam, 1995). Maximum Ni(II) removal of 53.65 mg/g took place at pH ranging from 4-5. In another column studies, Periasamy and Namasivayam (1996) explored Cu(II) removal from synthetic solution using peanut hull. Maximum Cu(II) removal of 65.57 mg/g occurred at pH ranging from 6-10. This result is significantly higher than that of Brown et al. (2000) which also employed peanut hulls in batch studies for Cu(II) removal with an adsorption capacity of 10.17 mg/g. This could be due to the inherent difference in the nature of both studies. In batch experiments, the concentration gradient decreased with an increasing contact time; while in column operation, the adsorbent continuously had physicochemical contact with fresh feeding solution at the interface of the adsorption zone, as the adsorbate solution passed through the column. Consequently, the metal removal by peanut hull in column studies was higher than that in batch studies (Rao et al., 2002).

Overall, waste from agricultural sources has demonstrated its ability for heavy metal removal. Chemical modification with an oxidizing agent can enlarge its surface area, thus improving its sorptive capacity. This process requires extra operational cost due to the consumption of chemicals. However, improved adsorption capacities of the adsorbents may compensate for the cost of additional processing. Moreover, when an adsorbent from agricultural waste get saturated, regeneration of the spent adsorbent has become one of the most cost effective and sound environmental option, since no solid waste will be generated and disposed of. In fact, regeneration and desorption could be done to recover valuable metal from the spent adsorbent. However, the economics of this practice are dependent on the characteristics of the regenerated carbon and on the mass losses incurred during regeneration. Compared to adsorbents from other sources, those from agricultural waste possess unique characteristics such as ease of regeneration and desorption with basic or acid solutions. This can be due to the fact that they have surface functional groups such as hydroxyl and carboxylic that possesses high affinity for metal cations.

Among the adsorbents derived from agricultural waste, hazelnut shell and soybean hull have demonstrated an outstanding metal removal for Cr(VI) (170 mg/g) and/or Ni(II) (154.9 mg/g). These adsorbents can be employed to treat inorganic effluent of concentration less than 100 mg/L, in the range of 100-1,000 mg/L and higher than 1,000 mg/L, providing one of promising alternatives to replace costly activated carbon. An acidic pH ranging from 2 to 6 is effective for metal removal by adsorbents from agricultural waste. The

Langmuir isotherm is generally applicable for the equilibrium data of such adsorbents, suggesting monolayer adsorption.

2.2.7.2. Industrial by-Product

Like agricultural waste, industrial by-products such as fly ash (Mavros et al., 1993), waste iron (Lee et al., 2003), metallic iron (Singh and Singh, 2003), hydrous titanium oxide (Ghosh et al., 2003), are inexpensive and abundantly available. They can be chemically modified to enhance its removal performance (Srivastava et al., 1989). However, unlike those from agricultural waste, adsorbents from this source can be obtained from industrial processing only.

Several studies have been conducted on the potential of industrial by-products for metal removal. Zouboulis et al. (1993) investigated red mud, a solid by-product from alumina production, for Ni(II) removal from simulated wastewater. Red mud had high cation exchange capacity (CEC) and cation exchange sites. With an initial Ni(II) concentration of 400 mg/L, the maximum Ni(II) uptake of 160 mg/g occurred at pH 9.0, where chemical precipitation of Ni(II) might also occur.

Lee et al. (2004) studied green sands, another by-product from the gray iron foundry industry, for Zn(II) removal from synthetic solution. Kinetics studies showed that the adsorbent was effective for the metal removal with an adsorption capacity of 32.46 mg/g. Clay and iron in the green sands played major roles for Zn(II) sorption via sorption and precipitation. The adsorption reaction was endothermic, as shown by the positive value of enthalpy change.

Blast-furnace slag is an adsorbent generated from steel plants. In 1996, Dimitrova studied the adsorption of Cu(II), Ni(II) and Zn(II) ions from simulated wastewater using blast-furnace slag. Its sorption capacities on Cu(II), Ni(II) and Zn(II) were 133.35 mg/g, 55.76 mg/g and 103.33 mg/g, respectively. The metal sorption was in the form of hydro-oxo complexes and the high sorption capacity was due to the formation of soluble compounds on the internal surface of adsorbent. The Freundlich isotherm was representative for metals adsorption.

Feng et al. (2004) investigated Cu(II) and Pb(II) removal from simulated wastewater using iron slag and/or steel slag. A pH range from 3.5 to 8.5 [for Cu(II)] and from 5.2 to 8.5 [for Pb(II))] was optimum for both iron and steels slags to maximize metal removal. The sorption of both metals for the two slags corresponded well with the Langmuir isotherm.

Magnetite (Fe_3O_4), the main component of converter slag in the steel industry, was explored for Ni(II) removal from simulated wastewater (Ortiz et al., 2001). The adsorption rates increased with initial metal concentration, but decreased with increasing temperature. The maximum Ni(II) adsorption of 18.43 mg/g occurred at pH ranging from 5.5-6.0. The adsorption equilibrium followed the Freundlich isotherm.

Activated slag, an adsorbent from thermal power plant with a commercial price of US$ 0.038/kg, was studied for the treatment of synthetic and real wastewater (Gupta, 1998). Using simulated solution (in batch studies), the adsorption capacities of activated slag were 30 mg/g for Ni(II) and 29 mg/g for Cu(II) ions. Using real wastewater (in column study), the capacities of activated slag were 66 mg/g for Ni(II) and 38 mg/g for Cu(II).

Gupta et al. (2003) explored bagasse fly ash, a solid waste from sugar industry, for Cd(II) and Ni(II) removal from synthetic solution. About 90% of 14 mg/L of Cd(II) and 12 mg/L of Ni(II) were removed at pH ranging from 6.0 to 6.5. The metal removal improved with an increasing temperature, suggesting an endothermic reaction.

Ni(II) and Cu(II) removal from simulated wastewater using Seyitomer and Afsin-Elbistain fly ash was evaluated by Bayat (2002). In Turkey, fly ash is a waste from thermal power plants. The adsorption capacities of Seyitomer fly ash for the two heavy metals were comparable to those of Afsin-Elbistain fly ash. As indicated by the applicability of the Langmuir isotherm for the equilibrium data of both metals, monolayer adsorption might occur on the surface of the adsorbent.

As a whole, adsorbents from industrial by-products such as iron/steel slag have some advantages for metal removal. In addition to having a wide pH range (pH 1.5-9.0) effective for metal removal, such adsorbents can be employed to treat inorganic effluent with metal concentration of less than 20 mg/L, in the range of 20-100 mg/L, or higher than 100 mg/L. Unlike adsorbents from agricultural waste, no further work has been done on the regeneration of the spent adsorbent from industrial by-products. This could be due to the fact that adsorbents derived from agricultural waste and industrial by-products have different characteristics in nature from each other such as processing conditions, making them difficult for regeneration. As a result, further R&D work needs to be undertaken to enable the adsorbents for commercialization.

Among the adsorbents from industrial by-products, red mud and blast-furnace slag have high metal adsorption capacities (160 mg Ni(II)/g and 133.35 mg Cu(II)/g), respectively. Their capacities were comparable to that of orange peel (158 mg Ni(II)/g) and of soybean hull (154.9 mg Cu(II)/g). However, unlike the two prominent adsorbents, fly ash, a by-product from lignite-burning plants, is not effective for metal removal due to its low metal-binding capacity. Both the Langmuir and the Freundlich isotherms can be applicable for the equilibrium data of adsorbents from industrial by-products, suggesting that either monolayer or multilayer adsorption could occur on their surface, depending on the type of by-products.

2.2.7.3. Natural Materials

Like industrial by-product, natural materials locally available in certain region can be employed as low-cost adsorbents. Due to its metal-binding capacity, natural materials such as zeolite and clay have been explored for treating metal-contaminated wastewater.

A. Zeolite

Among minerals that possess sorbent properties, zeolite appears as one of the most promising for metal purification. Zeolite, an alumino-silicate tetrahedron connected with oxygen atom, has charge-balancing cations that counter-balance the negative charge localized on the aluminosilicate framework (Keane, 1998). In Greece, zeolite has been used for the treatment of metal-contaminated wastewater due to its low cost. Depending on its quality, zeolite is sold at US$ 0.03-0.12/kg (Virta, 2001).

Babel and Kurniawan (2003) studied Cr(VI) uptake from simulated wastewater using natural zeolite. NaCl-treated zeolite had better removal capabilities (3.23 mg/g) for Cr(VI) ions than as-received zeolite (1.79 mg/g) at an initial Cr concentration of 20 mg/L. These results suggest that the Cr adsorption capacities of zeolite varied, depending on the extent of chemical treatment (Wingenfelder et al., 2005). The results were significantly lower than those of Perić et al. (2004), which also used zeolite for Zn(II) and Cu(II) removal (Zn(II): 13.06 mg/g; Cu(II): 25.04 mg/g). The metal removal by zeolite was a complex process, involving ion exchange and adsorption.

When zeolite in the column get exhausted, recovery of the adsorbed material as well as the regeneration of adsorbent becomes necessary. Kurniawan (2002) reported that NaCl was suitable to reuse zeolite with regeneration efficiency of more than 90%. This could be due to the fact that Cr adsorption on zeolite mostly occurred due to ion exchange between Na(I) of zeolite and Cr(III) ions in the solution.

B. Clay

Another low-cost mineral that has high cation exchange capacity (CEC) in solution is clay. There are three types of clay: montmorillonite, bentonite and kaolinite. Out of the three, montmorillonite has the highest CEC and its current market price (US$ 0.04-0.12/kg) is twenty-times cheaper than that of activated carbon (Virta, 2002).

A number of studies on metal uptake using montmorillonite have been conducted. The adsorption of Cd(II), Cr(III), Cu(II), Ni(II) and Zn(II) ions on Na-montmorillonite using column operations was investigated by Abollino et al. (2003). Cu(II) was the least adsorbed of the cations, while Cd(II) ion was the most adsorbed due to its high charge density. Modification of montmorrilonite using sodium dodecyl sulfate (SDS) was reported to significantly improve its removal for Cu(II) and Zn(II) (Lin and Juang, 2002). The amount of Cu(II) sorbed was larger (16.13 mg/g) than that of Zn(II) (13.27 mg/g). These results were four times higher than those of montmorrilonite modified with sodium (Abollino et al., 2003). This could be due to the fact that large anionic species such as SDS could easily enter and became fixed strongly in the interlayer region of montmorillonite, compared to NaOH.

Vengris et al. (2001) reported that pretreatment of clay with HCl significantly improved the removal of Ni(II), Cu(II) and Zn(II) from simulated wastewater. This could be due to the fact that acid treatment changed the chemical composition and mineralogical structure of clay, enhancing its uptake capacity. The maximum adsorption capacities for the cations in the solution were in the order: Cu(II) > Ni(II) > Zn(II). As indicated by the applicability of the Langmuir isotherm for the equilibrium data of those metals, monolayer adsorption might occur on the surface of adsorbent.

Another material from clay mineral to adsorb metal is bentonite, which consists of clay, silt and sand. This material is valuable for its tendency to absorb water in the interlayer site (Kaya and Oren, 2005). Chakir et al. (2002) studied the uptake of Cr(III) from simulated solution using bentonite and an expanded perlite. Surface complexation played major roles in Cr(III) removal. Kinetics studies showed that the Cr(III) uptake by bentonite was faster than that by perlite. The Cr(III) removal by bentonite (96%) was remarkably higher than that by perlite (40%) at the same Cr concentration of 20 mg/L. Similar results were also obtained by Ulmanu et al. (2002), who compared Cu(II) and Cd(II) removal from synthetic solution using bentonite. Compared to diatomite, bentonite has higher adsorption capacities for both ions (Cu(II): 18.16 mg/g; Cd(II): 9.34 mg/g).

Kaolinite is the other clay, representing a layered alumino-silicate mineral with the structure of a tetrahedral (Si center) and octahedral (Al center). Kaolinite was used for Cu(II), Ni(II), Mn(II) and Co(II) removal from simulated solution (Arias et al., 2002; Yavuz et al., 2003). The metal sorption on kaolinite followed the Langmuir isotherm. Due to their small ionic radius, Cu(II) ions had the highest adsorption affinities compared to other.

Chantawong et al. (2003) studied the adsorption of Cd(II), Cr(III), Cu(II), Ni(II) and Zn(II), on kaolinite and ballclay from simulated solution. Metal adsorption by kaolinite was observed as follows: Cr > Zn > Cu ≈ Cd > Ni > Pb and that by ballclay was: Cr > Zn > Cu > Cd > Ni. Ballclay has higher removal efficiency than kaolinite because illite, the major mineral in the ballclay, has a higher surface charge than kaolinite. Cr(III) ions were the most readily absorbed of the metals, as it had the highest ionic charge compared to the others.

Smith et al. (1996) investigated the sorption of Cd(II) and Zn(II) ions from simulated wastewater using glauconite, a complex clay mineral occurring in marine sediments. Equilibrium was attained within 60 min and Cd(II) ions had a higher adsorption capacity (4.1 mg/g) than Zn(II) (1.37 mg/g). Although the adsorbent is a complex mineral, the Langmuir isotherm was applicable for the equilibrium data.

Other natural materials such as pyrite and vermiculite have been studied less intensively due to its local availability. The Cr(VI) removal from simulated wastewater using pyrite fines was investigated (Zouboulis et al., 1995). Pyrite is a common mineral associated with sulphide ores and coal. Maximum Cr(VI) removal of 10 mg/g was achieved at pH ranging from 5.5 to 6.5. At higher pH than 9.0, $Cr(OH)_4^-$ species were formed, reducing Cr(VI) removal due to interference effects.

Álvarez-Ayuso and García-Sánchez (2003) also studied Ni(II) and Cu(II) removal from synthetic solution and/or real wastewater using natural vermiculite (NV) and exfoliated vermiculite (EF). In batch studies using simulated solution, maximum metal removal occurred at pH ranging from 5 to 6. The Langmuir isotherm was applicable for the equilibrium data. In column operations using real wastewater, at 218 bed volumes (BV), the Ni(II) concentration in the effluent was less than 2 mg/L, the discharge limit established by the European Community (EC).

The applicability of wollastonite for Cr(VI) uptake from synthetic wastewater was investigated by Sharma (2003). Maximum Cr adsorption of 0.52 mg/g occurred at pH 2.5. The metal adsorption was an endothermic reaction, as the removal increased with an increasing temperature from 30-50°C.

Overall, zeolite and clay have shown their ability to treat inorganic effluent. Among clay minerals, montmorrilonite and bentonite have shown reasonable removal for Cu(II) with adsorption capacities of 16.13 mg/g and 18.16 mg/g, respectively. The Langmuir isotherm is generally applicable to describe monolayer adsorption on the surface of adsorbent. Adsorbents derived from natural materials perform well at an acidic pH ranging from 2.5 to 6.5. Montmorrilonite and kaolinite generally treat inorganic effluent with metal concentration in the range of 100-1,000 mg/L or higher than 1,000 mg/L; zeolite and bentonite, however, are usually employed for treating real wastewater with metal concentration of less than 200 mg/L. Compared to that of adsorbents derived from agricultural waste, the price of those originated from natural materials is relatively higher, making them not competitive enough for commercial application.

2.2.7.4. Miscellaneous Low-Cost Adsorbents

Other adsorbents such as coal, sludge and peat have been studied less extensively due to its local availability. Karabulut et al. (2000) studied Cu(II) and Zn(II) removal from simulated wastewater with low-rank Turkish coals. Carboxylic acid and hydroxyl functional groups present on the surface of coal were the adsorption sites for metal removal via ion exchange. Maximum Cu(II) removal of 1.62 mg/g and 1.2 mg/g for Zn(II) occurred at pH 4.0. Despite using coal as the adsorbent, the metal capacities in the study carried out by Karabulut et al. were significantly lower than those of study undertaken by Solé et al. (2003) for Zn(II) removal with an adsorption capacity of 27.2 mg/g at pH 6.0, indicating that the adsorption capacity of an adsorbent depends on the initial concentration of adsorbate.

Sludge is another adsorbent used for Cu(II) and Cd(II) removal from simulated waste-water (Calace et al., 2003). The pH breakthrough curve of Cd(II) was similar to that of Cu(II) ions at pH 4.0. The sorption capacities of Cu(II) and Cd(II) ions were 9.5 and 10.23 mg/g, respectively. It is interesting to note that about 70% regeneration efficiency was achieved for both cations with 0.1 M HCl.

Ni(II) removal from synthetic solution using a combined sludge-ash was also studied (Weng, 2002). As indicated by the applicability of the Freundlich isotherm for equilibrium data, multilayer adsorption might occur on the surface of the adsorbent. The adsorption capacity of the combined adsorbent was 0.32 mg/g, lower than that of fly ash alone (0.65 mg/g) in another study (Weng, 1990). These results were very low compared to those of Zhai et al. (2004), who investigated Cd(II) and Ni(II) removal from simulated wastewater using sewage sludge. Zhai et al. reported that maximum removal of Cd(II) was 16 mg/g and of Ni(II) was 9 mg/g at pH ranging from 5.5 to 6.0, indicating that the characteristics of the individual adsorbent affect the removal capacity of an adsorbent on heavy metal.

Dean et al. (1999) investigated the uptake of Cr(III) and Cr(VI) ions from synthetic solution using peat. Maximum Cr(III) uptake of 14.04 mg/g took place at pH 4.0, while the optimum Cr(VI) uptake of 30.16 mg/g occurred at pH 2.0. Although the oxyanion of $Cr_2O_7^{2-}$ was used as the source of Cr(VI) in the wastewater, under common environmental pH (pH < 6.0), the most dominant species at pH 2.0 and pH 4.0 for metal removal was $HCrO_4^-$ (Lalvani et al., 1998; Pradhan et al., 1999). The hydrolysis reaction of $Cr_2O_7^{2-}$ was reported as follows:

$$Cr_2O_7^{2-} + H_2O \leftrightarrow 2\ HCrO_4^- \qquad pK_3 = 14.56 \qquad (2.12)$$

Since the acidic functional groups including carboxylic, hydroxyl, and carbonyl groups are present on the surface of peat, the physicochemical interactions that might occur during Cr(VI) removal could be expressed as follows:

$$M^{n+} + n\ (\text{-COOH}) \quad \leftrightarrow \quad (\text{-COO})_n\ M + n\ H^+ \qquad (2.13)$$

where (–COOH) represents the surface functional group of peat and n is the coefficient of the reaction component, depending on the oxidation state of metal ions, while the M^{n+} and the H^+ are Cr(III) and hydrogen ions, respectively.

Boonamnuayvitaya et al. (2004) investigated the removal of Cd(II), Cu(II), Zn(II) and Ni(II) ions from synthetic solution using pyrolized coffee residue and clay. The clay was used as a binder for the residue. Adsorption was observed in the order of Cd(II) > Cu(II) > Zn(II) > Ni(II). The Langmuir isotherm was representative for the equilibrium data of the metals, suggesting that monolayer adsorption might occur on the surface of adsorbent.

Gang et al. (2000) studied the Cr(VI) removal from simulated wastewater using a reactive polymer, a long alkyl quaternized poly 4-vinylpyridine (PVP). Metal adsorption by the adsorbent was strongly influenced by pH. Maximum Cr(VI) of 3.5 mg/g occurred at a pH range of 4.5-5.5. The Langmuir isotherm was representative for the equilibrium data, indicating monolayer adsorption on the surface of the adsorbent.

Among the adsorbents presented above, peat and wool stand out for high Cr(VI) removal with adsorption capacities of 30.16 mg/g and 41.15 mg/g, respectively. An acidic pH range of 2.0-4.0 is effective for Cr(VI) removal with an initial metal concentration ranging from 100 mg/L to 200 mg/L. It is important to note that the adsorption capacities of low-cost adsorbents presented above vary, depending on the characteristics of the individual adsorbent, the extent of surface modification and the initial concentration of adsorbate.

2.2.7.5. Activated Carbon (AC)

Based on its size and shape, there are four types of activated carbon: powder (PAC), granular (GAC), fibrous (ACF), and cloth (ACC). Due to the different raw materials, the extent of chemical activation, and the physicochemical characteristics; activated carbon has its specific application as well as inherent advantages and disadvantages in wastewater treatment. Depending on its quality, activated carbon is costly at US$9/kg (CSIRO, 2004).

Netzer and Hughes (1984) studied the adsorption of Cu(II) ions by granular activated carbon (GAC) type Barney Cheney NL 1266 from synthetic solution. At pH 4.0, GAC could remove 93% of 10 mg/L Cu(II) solution. The results were lower than that of GAC type Filtrasorb 400, employed by Sharma and Forster (1996) for Cr(VI) removal from simulated wastewater. They reported that the maximum metal adsorption capacity of 145 mg/g was achieved at pH ranging from 2.5 to 3.0.

The adsorption of Cd(II) ions from synthetic solution using GAC type Filtrasorb 400 was investigated (Leyva-Ramos et al., 1997). The metal adsorption capacity of GAC was 8 mg/g at pH 8.0. This result was slightly higher than that of Bansode et al. (2003) which employed GAC type Filtrasorb 200 for Cu(II) removal with an adsorption capacity of 6.10 mg/g, indicating that the initial concentration of adsorbate plays major roles in determining the extent of the adsorption capacity of an adsorbent. For Zn(II) removal, the adsorption capacity of GAC type C (Leyva-Ramos et al., 2002) was significantly higher than that of GAC type Filtrasorb 200 (Bansode et al., 2003).

To improve its removal performance on Cr(III) ions, the surface of GAC was chemically modified with nitric acid (Aggarwal et al., 1999). The adsorption capacity on the oxidized carbon for Cr(III) was enhanced three times to 30 mg/g. This could be due to the greater negative charge on the surface of oxidized GAC compared to the non-oxidized material (Goel et al., 2005). Due to electrostatic attraction between the Cr(III) ions and the negatively-charged surface, more adsorption of Cr(III) ions might occur on the carbon surface, resulting in higher metal uptake by the adsorbent. Similar results were obtained by Shim et al. (2001), Park and Jung (2001), Rangel-Mendez and Streat (2002), Monser and Adhoum (2002), Babić et al. (2002), and Babel and Kurniawan (2004a). They also found that electrostatic attraction played a major role in metal adsorption on the surface of carbon.

Overall, chemical modification of activated carbon can improve its removal for heavy metal. Among various type of activated carbon, GAC type Filtrasorb 400 and HNO_3-oxidized activated carbon fiber (ACF) stand out high metal removal with adsorption capacities of 145 mg of Cr(VI)/g and 146 mg of Cd(II)/g, respectively. The results were lower than those of adsorbent originated from agricultural waste such as hazelnut shell (Cr(VI): 170 mg/g) and citric-acid modified soybean hull (Cu(II): 154.9 mg/g). It is interesting to note that activated carbon performs effectively at an acidic pH range of 2.5-7.0 and has the ability to treat inorganic effluent with metal concentration ranging from 10-1,000 mg/L. Due to its high cost, industrial users are reluctant to employ activated carbon for the treatment of metal-contaminated wastewater, opening the way for commercialization of low cost adsorbent derived from agricultural waste (Inbaraj and Sulochana, 2004).

2.3. COMPARISON OF ADSORPTION CAPACITIES AMONG LOW-COST ADSORBENTS AND AC

To justify their viability as effective adsorbents for heavy metal uptake, the adsorption capacities of all low-cost adsorbents need to be compared. Kurniawan et al. (2006) reviewed over 193 published articles (1977-2005) on the application of various low-cost adsorbents derived from agricultural waste, industrial by-product or natural material for the removal of heavy metals (Cd(II), Cr(III), Cr(VI), Cu(II), Ni(II) and Zn(II)) from metal-contaminated wastewater.

It is evident from the literature survey that low cost adsorbents from agricultural waste that have demonstrated outstanding metal removal capabilities can be viable alternatives to costly activated carbon for the treatment of metals-contaminated wastewater. The removal capacity of such materials (Cr(VI): 170 mg/g of hazelnut shell, Ni(II): 158 mg/g of orange peel, Cu(II): 154.9 mg/g of soybean hull, Cd(II): 52.08 mg/g of jackfruit) is comparable to that of activated carbon (Cd(II): 146 mg/g, Cr(VI): 145 mg/g, Cr(III): 30 mg/g, Zn(II): 20 mg/g), suggesting the viability of low-cost adsorbents for treating metals-contaminated wastewater.

Among adsorbents derived from agricultural waste, hazelnut shell (Cr(VI): 170 mg/g), orange peel (Ni(II): 158 mg/g) and citric acid-modified soybean hull (Cu(II): 154.9 mg/g) stand out for significantly higher metal adsorption capacities, compared to those from natural material such as clay (Ni(II): 81 mg/g, Cu(II): 83 mg/g, Zn(II): 63 mg/g). This can be due to the fact that most of the adsorbents originated from agricultural waste contain cellulose, hemicellulose and lignin, with functional groups such as carboxylic, carbonyl and hydroxyl, which possess high affinity for heavy metal ions. Tan et al. (1993) reported that metal uptake by adsorbents from agricultural waste might occur through sorption process involving these surface functional groups.

An objective evaluation of the commercial potential of any material as an adsorbent should consider its local availability, as this factor is closely related with cost minimization (Bailey et al., 1999). For countries not producing agricultural waste, the cost of transportation of raw material may be involved for the production of activated carbon. Local environmental conditions also affect the removal performance of activated carbon derived from agricultural waste. In tropical countries, activated carbon shows a favorable removal performance, as adsorption occurs at higher temperature, resulting in higher metal uptake.

Different characteristics of individual adsorbents can be addressed in different ways to improve its metal removal performance. Low cost adsorbents from agricultural waste can be chemically modified with oxidizing agents to enhance their metal removal performance. This can be due to the increased surface area for metal adsorption and greater negative surface charge, resulting from oxidative treatment. The adsorption capacity is raised through columbic attractions with metal in the solution (Babel and Kurniawan, 2004b). However, those from natural materials such as zeolite need to be treated with NaCl to improve its metal removal performance. Kurniawan (2002) reported that this treatment rendered the adsorbent in the homoionic form of Na(I). Consequently, the Na(I) of zeolite could be replaced by Cr(III) ions in the solution through ion exchange process (Kurniawan, 2002).

2.4. COST-EFFECTIVENESS OF LOW-COST ADSORBENTS

Since the cost-effectiveness of an adsorbent is one of the important issues that must be considered when selecting an adsorbent, the price of low-cost adsorbents has to be compared. The price of low-cost adsorbents is not presented due to the unavailability of data from previous studies. Expenditure on individual adsorbents varies, depending on the processing employed and its local availability (Hasan et al., 2000; Shukla and Pai, 2005).

Although expenditure on low-cost adsorbents may be negligible, further cost-benefit analysis needs to take into account any spending associated with regeneration or operation including chemicals, electricity, labor, transportation and maintenance (Krishnan and Anirudhan, 2003). If the overall cost remains economically attractive, low-cost adsorbents for metal uptake present promising alternatives to costly activated carbon, as their use can minimize treatment cost (Kurniawan et al, 2006a).

The cost effectiveness of outstanding adsorbents such as hazelnut shell for metal removal from wastewater increases if they can be regenerated for multiple uses. The price of low-cost adsorbents may decrease, as more industries consider their use for treating metals-contaminated wastewater. Moreover, local availability of low-cost adsorbents, fast adsorption rate, reasonable adsorption capacity and sludge-free operation will reduce operational cost and render their use for treating inorganic effluent is a very attractive option. In general, technical applicability and cost-effectiveness are the key factors that play major roles in the selection of the most suitable adsorbent to treat inorganic effluent (Kurniawan et al, 2006b).

In spite of the positive trend of the development worldwide, so far no low cost adsorbents have been available for commercialization. This indicates that more R&D work and in-depth study on the mass production of the adsorbents with enhanced removal performance and on the most suitable chemicals for their regeneration, need to be addressed prior to commercialization.

2.5. EVALUATION OF HEAVY METAL REMOVAL BY DIFFERENT PHYSICO-CHEMICAL TREATMENTS

To evaluate the performances of all the treatments described above, a comparative study is presented in terms of pH, dose required (g/L), initial metal concentration (mg/L), and metal removal efficiency. Although it has a relative meaning due to the different testing conditions (pH, temperature and strength of wastewater), this comparison is useful to evaluate the overall removal performance of each treatment in the decision-making process.

Depending on the initial metal concentration of the contaminated wastewater, it is evident from the table that ion exchange has achieved a complete removal of Cd(II), Cr(III), Cu(II), Ni(II) and Zn(II) with an initial concentration of 100 mg/L, respectively. The results are comparable to that of reverse osmosis (99% of Cd(II) rejection with an initial concentration of 200 mg/L). Lime precipitation has been found as one of the most effective means to treat inorganic effluent with a metal concentration of higher than 1,000 mg/L. It is interesting to note that flotation also offers comparable metal removal (100%) to RO, an advanced treatment technique, at the same metal concentration of 50 mg/L, but with a lower cost.

In general, physico-chemical treatments offer various advantages such as their rapid process, ease of operation and control, flexibility to change of temperature. Unlike biological system, physico-chemical treatment can accommodate variable input loads and flow such as seasonal flows and complex discharge. Whenever it is required, chemical plants can be modified. In addition, the treatment system requires a lower space and installation cost. Their benefits, however, are outweighed by a number of drawbacks such as their high operational costs due to the chemicals used, high energy consumption and handling costs for sludge disposal. However, with reduced chemical costs and a feasible sludge disposal, physico-chemical treatments have been found as one of the most suitable treatments for inorganic effluent.

2.6. COST COMPARISON OF TREATMENT TECHNOLOGIES FOR ELECTROPLATING WASTEWATER

To estimate a reliable treatment cost for metal-contaminated wastewater is difficult due to the many cost components such as pumping equipment and treatment facility involved. In addition, changes in the quality and quantity of the plating wastewater due to the fluctuating market demand also contribute to the variations of its treatment cost. Therefore, information on the treatment cost of plating wastewater is rarely reported.

Basically, the treatment cost of contaminated wastewater varies, depending on its strength and quantity, the process employed, the amount and composition of impurities, as well as the extent of purification (Weber and Holz, 1992; Hurd, 1999). The overall treatment cost includes the construction costs as well as the operational and maintenance costs (O&M). The construction costs normally depend on the effluent quality required and the capacity of the installation, while the O&M costs cover manpower, energy, chemicals and maintenance. The manpower cost varies from one country to another. To obtain an accurate assessment of the operational cost for the treatment of electroplating wastewater, a pilot-scale study needs to be carried out (Tsagarakis, 2003).

Although this article has featured some pilot studies (Wong et al, 2002; Martinez et al, 2004), most of the data presented above are derived from research conducted on a laboratory scale. Consequently, further experiments on a pilot scale are needed to quantify the overall treatment cost associated with the proposed treatment. A direct comparison of the overall treatment cost of each technique presented above may be difficult due to their different operating conditions.

In addition, the overall treatment cost for electroplating wastewater varies depending on the process employed and the local conditions. This may be attributed to the fact that most of the electroplating industries (especially in the Southeast Asia) are located either in the commercial area of a town or in industrial estates, where wastewater is discharged into sewers after neutralization with acids/alkalis (Lo and Tao, 1997). Wastewater treatment plants of other engineering industries that have electroplating sections are designed to handle the entire wastewater, including those from the electroplating process. As a result, the cost of installation and operation of such plants do not represent the actual cost of an independent plating industry. Wide variations in the flow and the characteristics of the effluent wastewater also present difficulties in estimating the treatment cost accurately. Such inconsistency in data presentation makes a cost comparison among the available treatment technologies for wastewater laden with heavy metals difficult to materialize.

CONCLUSION

Over the past two decades, environmental regulations have become more stringent, requiring an improved quality of treated effluent. In recent years, a wide range of treatment technologies such as chemical precipitation, coagulation-flocculation, flotation, ion exchange and membrane filtration, have been developed for heavy metal removal from contaminated wastewater. It is evident from the literature survey of 193 articles (1977-2005) that ion exchange and membrane filtration are the most frequently studied and widely applied for the treatment of metal-contaminated wastewater. Ion exchange has achieved a complete removal of Cd(II), Cr(III), Cu(II), Ni(II) and Zn(II) with an initial concentration of 100 mg/L, respectively. The results are comparable to that of reverse osmosis (99% of Cd(II) rejection with an initial concentration of 200 mg/L). Lime precipitation has been found to be effective to treat inorganic effluent with a metal concentration higher than 1,000 mg/L.

In addition, low-cost adsorbents from agricultural waste have demonstrated outstanding removal capability (Cr(VI): 170 mg/g of hazelnut shell, Ni(II): 158 mg/g of orange peel, Cu(II): 154.9 mg/g of citric-acid modified soybean hull, Cd(II): 52.08 mg/g of jackfruit) compared to activated carbon (Cd(II): 146 mg/g, Cr(VI): 145 mg/g, Cr(III): 30 mg/g, Zn(II): 20 mg/g). It is important to note that the metal adsorption capacities of low-cost adsorbents presented above vary, depending on the characteristics of the individual adsorbent, the extent of surface modification and the initial concentration of adsorbate.

Although many techniques can be employed for the treatment of wastewater laden with heavy metals, it is important to note that the selection of the most suitable treatment for metal-contaminated wastewater depends on the initial metal concentration, the overall treatment performance compared to other technologies, plant flexibility and reliability, environmental impact as well as economics parameter such as the capital investment and operational costs (energy consumption and maintenance). Finally, technical applicability, plant simplicity and cost-effectiveness are the key factors that play major roles in the selection of the most suitable treatment system for inorganic effluent. All the factors mentioned above should be taken into consideration in selecting the most effective and inexpensive treatment in order to protect the environment.

Chapter 3

ANALYTICAL METHODOLOGY

ABSTRACT

This chapter presents and describes pretreatment methods of surface modification and standard tests that were conducted to determine Cr(VI) concentrations after adsorption treatment using various types of low cost adsorbents and commercial activated carbon (CAC) in batch and column modes. Each section in this chapter elaborates chemical reagents and standard procedures that will be employed not only to characterize the physico-chemical characteristics of real wastewater samples collected from local electroplating industries in Rangsit (Thailand), but also to determine Cr(VI) concentrations after adsorption treatment. All the tests employed in this study were adopted from the *Standard Methods for the Examination of Water and Wastewater* (1998).

3.1. MATERIAL AND METHODS

Unless otherwise stated, all the chemicals and reagents were of analytical grade supplied by Merck (US) and used without further purifications. Deionized water was employed to prepare all the working solutions and reagents. The adsorbents used in the present study are commercial grade. All glassware was soaked for 2 h in tap water and then rinsed with deionised water. Before being used, they were dried in an oven and cooled.

3.1.1. Synthetic Chromium Wastewater

Potassium dichromate ($K_2Cr_2O_7$) was used as the source of Cr(VI) in synthetic wastewater (Sharma and Forster, 1996; Gupta et al., 1999; Bayat, 2002). Before being dissolved in deionized water, $K_2Cr_2O_7$ in the fine crystal form was dried in an oven at $100°C$ for 30 min and cooled in a desiccator at ambient temperature. The stock solution of 500 mg/L was prepared by dissolving 0.7071 g of $K_2Cr_2O_7$ in 500 mL of deionized water, while Cr concentration of synthetic wastewater was varied from 5 to 10 mg/L by diluting the stock solution. Before being used, the pH of Cr solution at 20 mg/L of Cr concentration was measured using a pH meter with an initial pH 5.2. Adjustment of pH was also conducted using 0.1 N NaOH and/or 0.1 N H_2SO_4 (Demirbaş et al., 2002; Selvaraj et al., 2003). Agitation of the system under investigation was carried out on a rotary shaker, while the remaining Cr concentrations after adsorption were analyzed using a spectrophotometer type CECIL/CE 1021 (Cambridge, England).

3.1.2. Real Chromium Wastewater from Electroplating Industry

The real Cr wastewater sample, collected from a local electroplating industry located in Rangsit (Thailand), was stored in 20 L polyethylene carboys that were filled to capacity and capped tightly until being used for batch and column studies. The samples were stored in a refrigerated storage chamber at $4°C$ to minimize any further changes that might occur in their physico-chemical properties prior to experiments. After their collection, the samples were immediately characterized according to the Standard Methods (1998) for the following parameters: pH, turbidity, total suspended solids, total dissolved solids, conductivity, hexavalent chromium, total chromium, COD, and BOD_5.

3.1.3. Chitosan

Chitosan flake 90% deacetylated, supplied by Eland. (Bangkok) was washed and dried.

3.1.4. Surface Modification of Adsorbent

The adsorbents used in present studies are CSC type 12/40 and CAC type PHO 8/35 LBD, supplied by Carbokarn (Bangkok, Thailand), as well as zeolite type commercial grade. Their physical characteristics are listed in Table 3.1.

Table 3.1. Physical properties of CSC, CAC, zeolite, and chitosan

Property	CSC (12/40)	CAC (PHO LBD 8/35)	Zeolite	Chitosan
Particle size (mm)	0.42	0.50	0.65	
Solid density (g/cm^3)	0.63	0.48	2.00	
Packing density(g/cm^3)	0.73	0.53	2.30	
Total surface area (m^2/g)	5-10	900-1100	780	
Pore volume (mL/g)	0.06	0.73		
Base material	Coconut shell	Coconut shell		Shrimp waste
Deacetylation degree (%)				$90 \pm 5\%$
Cation exchange capacity (meq/g)			2.20	
Unit price (Baht/kg) (US$/kg)	20 0.46	60 1.37	30 0.67	50 1.14

Remarks: 1US$= 43.72 Baht (in April 2003).

Both granular CSC and CAC in as-received form were chemically modified with strong oxidizing agents such as sulfuric and/or nitric acids, and coated with chitosan respectively, while zeolite alone was treated with sodium chloride. The complete method of chemical modifications/pretreatment for each individual adsorbent is listed as follows:

A. Oxidation of CSC with Sulfuric Acid

For conditioning, CSC was washed with deionized water until any leachable impurities due to free acid and adherent powder were removed. The samples were then treated with 2% H_2SO_4 (v/v) in an incubator at 110°C for 24 h and washed with deionised water until the pH of solution was stable. Afterwards, CSC was soaked in 2% $NaHCO_3$ (w/v) to remove any residual acid left. Finally, the sample was dried overnight in an oven at 110°C, cooled at room temperature, and stored in a desiccator until required for use (Kadirvelu et al., 2001; Selvi et al., 2001, Demirbaş et al., 2002).

B. Oxidation of CSC with Nitric Acid

After being washed with deionised water and dried overnight, CSC was oxidized with 65% HNO_3 (w/v). A known volume of the nitric acid was heated in an incubator at $110^{\circ}C$ for 3 h. The CSC was then immersed in it (volume ratio of HNO_3 and CSC= 5:1) and oxidized for 3 h. After cooling, the acid solution was drained and the oxidized CSC was washed with deionised water until the pH of rinsing water remained constant. Finally, the samples were dried overnight in an oven at $110^{\circ}C$ and stored in a desiccator (Lisovskii et al., 1997; Pittman et al., 1997; Pradhan and Sadlae, 1999b; Ramos et al., 2002; Pereira et al., 2003).

C. Coating CSC with Chitosan

To produce CSC coated with chitosan (CSCCC), chitosan flakes 90% deacetylated were diluted in 0.5% acetic acid (v/v). The diluted mixtures were mechanically agitated for 24 h to form a homogenized gel. Afterwards, CSC was dipped into the gel, where the ratio of chitosan and CSC dose = 1:5, and then gently shaken overnight. Afterwards, the gel-coated CSC was washed with deionized water and dried. This process was repeated until a thicker coating on CSC was formed. The coated activated beads were removed and neutralized by putting them in 0.5% NaOH (w/v) solution for 3 h. The chitosan gel beads were then extensively rinsed with deionized water and dried. Hereafter, it was called CSC coated with chitosan (non-treated CSCCC). Similarly, sulfuric-treated CSC was also coated with chitosan and this type of adsorbent was called sulfuric-treated CSC coated with chitosan (sulfuric-treated CSCCC).

D. Treatment of Zeolite with NaCl

Prior to experiments, the zeolite was treated with $2M$ NaCl. The suspension was continuously agitated for 24 h using a rotary shaker. Zeolite was then separated from the supernatant using GF/C filters and the liquid was drained. The washing process was repeated to remove any excess NaCl on the surface of the zeolite. Finally, the zeolite was dried for 3 h and stored until required for further use (Blanchard et al., 1984; Zamzow et al., 1990; Mondale et al., 1995; Ali and Bishtawi, 1997; Curkovic et al., 1997; Inglezakis et al., 1999, 2001).

E. Oxidative Treatment of Cac

The same procedures, as of CSC presented in section 3.1.4.a and 3.1.4.b, were also employed for the surface modification of CAC with sulfuric and nitric acids respectively.

3.1.5. Diphenylcarbazide Solution

To analyze the remaining Cr concentration in the samples after adsorption treatment, about 0.25 g of 1,5-diphenylcarbazide was dissolved in 50 mL of acetone and then stored in a brown bottle.

Table 3.2. List of equipment for batch and column studies

Type	Items	Model	City of origin
Analytical equipment	pH meter	pH PRO	Singapore
	Spectrophotometer	CECIL/CE 1021	Cambridge, England
	Conductivitimeter	TPS WP84	Brisbane, Australia
	Turbidimeter	HACH 2100P	Düsseldorf, Germany
Non-analytical equipment	Rotary shaker	Sseriker II PNP	Bangkok, Thailand
	Filtering apparatus	GF/C	Bangkok, Thailand
	Peristaltic pump	SP 311 VELP Scientifica	Milan, Italia
	Column	Glass Column	Bangkok, Thailand
	Oven	Memmert MEM 1-UE 500	Schawabach, Germany
	Desicator	Bossman DBK-78	Taipeh, Taiwan
	Incubator	VELP FOC-2251	Milan, Italia

3.2. EQUIPMENT

In order to accomplish the overall objective in this research, batch and column studies were designed to ensure that they were technically feasible, engineering applicable, highly efficient for Cr(VI) removal, simple and easy in terms of handling and controlling the operations. Therefore, the experimental equipment consists of analytical and non-analytical equipment, as presented in Table 3.2.

3.3. EXPERIMENTAL PROCEDURE

The overall experiments were divided into two consecutive phases: batch and column studies. Batch studies primarily dealt with the study of adsorption mechanism of Cr(VI) removal from electroplating wastewater by the adsorbents, while column studies were designed to develop a laboratory scale for the possibility of industrial applications. Generally, the design of the overall study can be illustrated by the flowchart below.

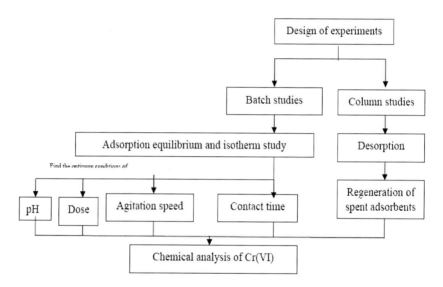

Figure 3.1. Design of batch and column studies.

3.3.1. Batch Experiments

Batch study is a convenient way to assess the removal capability of various low-cost adsorbents on Cr, as it yields valuable information on the capacity of media for the metal ions and the rate of metal uptake. This information would help in designing column operations by yielding data necessary for determining media usage rate, column size, and breakthrough time prediction. For this reason, batch studies were conducted to provide preliminary data of the adsorbent's effectiveness on Cr removal before running column operations.

All batch experiments, investigating the optimum condition of dose, pH, speed, and contact time (Chakravarty et al., 2002; Demirbaş et al., 2002; Selvaraj et al., 2003; Wong et al., 2003), were conducted using real Cr wastewater.

A. Effects of Dose

The removal efficiency of Cr(VI) varies with the dose of adsorbent. Therefore, a study determining the optimum dose for each adsorbent was conducted to achieve a maximum chromium adsorption during equilibrium conditions.

To find the effects of optimum dose on the removal efficiency of Cr(VI), the amount of dose was varied from 0.5 to 30 g/l. Equilibrium conditions were obtained by allowing the mixture to equilibrate under a constant agitation speed and at ambient temperature. At the designated time, the two phases under investigation were separated using GF/C filters and the supernatant was collected for chromium analysis using a spectrophotometer.

The amount of total Cr(VI) removed was taken as the concentration difference between the originally added and the finally remaining. The removal efficiency of adsorbent on Cr(VI) (%REA) was defined as:

$$\left(\%REA \right) = \frac{\left(c_0 - c_1 \right)}{c_0} \times 100 \tag{3.1}$$

where: C_0 and C_1 are the initial and final concentration of chromium (mg/L) respectively.

B. Effects of pH

Earlier studies have indicated that knowledge of the optimum pH is very important as pH is considered as the most influencing factors in the adsorption. The pH of solution affects not only the surface charge of the adsorbent, but also the degree of ionization and the speciation of adsorbate species. Therefore, this experimental work investigating the optimum pH was performed to ascertain its effects on the removal efficiency of the adsorbents on Cr(VI).

Initially a fixed Cr concentration was prepared from the stock solution. About 100 mL of Cr solution was taken into a number of 250 mL conical flasks and the optimum dose of adsorbent was added. The initial pH of samples was adjusted to a fixed value in the range of 2.0-9.0 using 0.1 N

H_2SO_4 and/or 0.1 N NaOH. The pH of solution was measured prior to the addition of the adsorbent and again measured at the end of reaction. The pH, giving the maximum Cr removal, was selected and utilized for the remaining adsorption studies.

C. Effects of Agitation Speed

Previous studies have reported that agitation speed influences the degree of physico-chemical interaction between adsorbent and adsorbate in the solution. Therefore, the optimum agitation speed in this study was investigated to maximize the interactions between target metal ions and the adsorption site of the adsorbent during equilibrium.

To investigate the effects of agitation speed on the Cr removal efficiency of adsorbent, the same treatment method described in section 3.3.1.b was also employed with six different agitation speeds of the rotary shaker (90, 110, 130, 150, 170, and 190 rpm).

D. Effects of Contact Time

The removal of Cr(VI) was also studied as a function of contact time at optimum conditions of dose, pH, and agitation speed. The rate at which adsorption takes place is an important factor when designing an adsorption system. Consequently, it is necessary to establish the time dependence of metal capture for maximizing the rate of Cr uptake by the adsorbent during equilibrium.

The effects of contact time were investigated by adding the optimum dose of adsorbent into 100 mL of Cr solution at optimum pH and agitation speed. Sample was then collected on the designated contact time. The Cr concentration, solution pH, agitation speed, and contact time used for all the batch experiments are specified in the figures with the error bars of Cr removal efficiency for all types of adsorbents of less than 1.0%.

E. Adsorption Isotherm Study

Before running column tests, adsorption isotherm studies have been traditionally employed for preliminary investigation, as they describe how adsorbate has physico-chemical interactions with adsorbent in equilibrium conditions. Therefore, isotherm studies were conducted to optimize the design of an adsorption system for metal removal from effluents.

In this research, adsorption isotherm studies were carried out in the reaction mixture consisting of 1.5 g/L of adsorbent and 100 mL of Cr solution with varying Cr concentration from 5 to 10 mg/L at the same conditions of pH, agitation speed, and contact time (Periasamy and Namasivayam, 1996; Rivera-Utrilla et al., 2003; Netpradit et al., 2003; Hasar, 2003).

To establish the most appropriate correlations for the equilibrium curve, data obtained from equilibrium studies were evaluated using the empirical equations of the Langmuir and Freundlich isotherms and then statistically compared. The coefficients of least-square fitting computed for the two isotherms indicated the existing relationship between the observed data at equilibrium conditions.

F. Effects of COD Removal

In order to produce water with sufficient quality for reuse, the initial COD value obtained from the real wastewater samples has to be substantially lowered after treatment with an adsorbent. Therefore, the COD value of the wastewater was determined to characterize the wastewater effluents from an electroplating industry.

The COD, an indication of the overall oxygen consumed during the oxidation of the oxidizable organic matter, is equal to the amount of dissolved oxygen that a sample will absorb from a hot acidic solution containing $K_2Cr_2O_7$ and $Hg(II)$ ions. Approximately 50 mL of the sample was refluxed with $0.04167M$ $K_2Cr_2O_7$ and H_2SO_4 in the presence of $HgSO_4$ and Ag_2SO_4 for 2 h at 150°C. The addition of $HgSO_4$ neutralized the effects of Cl^- ions present, while the latter acted as a catalyst for the oxidation. The excess of $K_2Cr_2O_7$ was titrated against $0.25M$ $Fe(NH_4)_2(SO_4)_2$ using ferroin as an indicator.

The amount of $K_2Cr_2O_7$ used corresponds to the oxidizable organic matter present in the sample as suggested by the following formula (Standard Methods, 1998):

$$COD(mg/l)= \frac{(b-a) \times (N) \, of \, ferrous \, ammonium \, sulfate \, (FAS) \times 8000}{Volume \, of \, sample \, (mL)} \quad (3.2)$$

where: b is the volume of FAS using blank (mL), while a represents the volume of FAS using sample (mL)

3.3.2. Column Experiments

Batch experiments were used for preliminary investigations and fixing the operational parameters, but in practice, technical systems normally use column operations to obtain a factual design model of the system.

In this study, column operations were conducted to validate the results of previous batch experiments and evaluate a flow-through system using real wastewater samples. By using continuous column operations, it was possible to determine the Cr adsorption capacity of each individual adsorbent and its technical feasibility for industrial applications.

A. Column Operation

In a fixed-bed study, a glass column, 50 cm of length with an internal diameter of 1.5 cm, was packed with a known mass of an adsorbent. At the bottom of the column, 1 cm of glass wool and glass beads layer was fitted.

The designated column was filled by the adsorbent. Feeding solutions containing Cr(VI) with a concentration of 20 mg/L was then prepared from the stock solution. After adjusting its pH to optimum based on the results of batch studies, the feeding solutions were introduced at the top of the column and accordingly, the column was operated with the feeding solutions flowing from top to bottom (down flow mode).

The pH of the effluents was recorded hourly to monitor any change that might occur during the adsorption. A flow rate of 6.0 mL/min was maintained with a peristaltic pump. This flow rate might slightly vary between the runs. The effluent samples were periodically collected by a fraction collector and then analyzed for residual metal concentrations.

Based on the wastewater discharge standard of the Thai Ministry of Natural Resources and Environment (2003), the Cr uptake at the 5% breakthrough point was selected to represent the operational capacity of the column. The column operations were terminated as soon as the saturation point was achieved and there was no difference between influent and effluent concentration ($C_e/C_o=1$). In this condition, all surface sites of the packing materials were occupied by adsorbed chromium. The column was then rinsed with deionised water to remove any unadsorbed Cr. These same methods were also employed for real Cr wastewater samples.

B. Column Regeneration

Regeneration of a spent adsorbent is necessary when the adsorbent used is expensive or if it is not always available in large quantities. From an economical point of view, an adsorbent can be considered to be efficient and effective if it is easily regenerated and re-utilized as frequently as possible without altering its removal performance on certain metal. Therefore, column regeneration using selected chemicals was undertaken to restore the removal performance of the same column to its original state.

After complete exhaustion (C_e/C_o=1), the column was desorbed by passing regenerants to recover the accumulated chromium on the spent adsorbent. A combination of $0.1M$ NaOH and HNO_3 solution was used as regenerants for both spent CSC and saturated CAC, while NaOH solution alone was used for the regeneration of spent zeolite. Desorption was terminated as soon as effluent metal concentration was negligible. Afterwards, the column was washed by deionised water at the same flow rate of 6.0 mL/min until the effluent pH was equal to its influent of 7.0-7.2. To quantify the regeneration efficiency (RE) of the spent adsorbent and evaluate its reusability, the method outlined by Martin and Ng (1987) was employed as follows:

$$\left(\%_{RE}\right) = \frac{\left(A_r\right)}{\left(A_0\right)} \times 100 \tag{3.3}$$

where Ao and Ar are the adsorption capacity of adsorbent before and after regeneration, while the percent loss in Cr adsorption capacity (%LAC) (that is the fraction of adsorbate that could no longer be adsorbed to that adsorbed during the first cycle) was calculated as:

$$\left(\%_{LAC}\right) = \frac{\left(A - I\right)}{A} \times 100 \tag{3.4}$$

where A and I represent the amount adsorbed during the first cycle and the amount adsorbed in each subsequent cycle (Shawabkeh et al., 2002).

3.3.3. Chemical Analysis of Chromium Concentration

Changes in the Cr(VI) concentrations in the solution after adsorption treatments were determined colorimetrically according to Standard Methods (1998). A purple-violet colored complex was generated in the reactions between Cr(VI) and 1,5-diphenylcarbazide in acidic condition. Absorbance was measured at wavelength (λ) 540 nm after 10 min of color development. The minimum detectable concentration by this method is 0.005 mg/L as Cr(VI). In acidic solutions, both $HCrO_4^-$ and $Cr_2O_7^{2-}$ anions can be detected.

Total chromium was determined by oxidizing Cr(III) with potassium permanganate in the presence of H_2SO_4 before reacting with 1,5-diphenylcarbazide. The difference between total Cr and Cr(VI) was the concentration of Cr(III).

3.4. STATISTICAL ANALYSIS

In order to ensure the accuracy, reliability, and reproducibility of the collected data, all the experiments were carried out in duplicate and the mean values of two data sets are presented. In most cases, the accuracy of the collected data was very good, as the relative standard deviation of Cr removal for all the experiments was found to be less than 1.0%. When the relative error exceeded this criterion, the data were discharged and a third experiment was conducted until the relative error fell within an acceptable range.

Statistical analysis using the paired t-test was performed to evaluate if there was any significant difference in terms of Cr removal efficiency for two categorical variables between an as-received adsorbent (before treatment) and a chemically modified adsorbent (after treatment) for CSC and/or CAC. This statistical test was conducted if the Kolmogorov-Smirnov test confirmed the normality of the variable distribution ($p > 0.05$), Otherwise, the Wilcoxon-signed rank test was performed for such a purpose.

Similarly, the analysis of variance (Anova) was employed to analyze data, which has three, or more categorical variables, if the distribution of each variable was normal ($p > 0.05$). Otherwise, Kruskall-Wallis test was employed.

One-way Anova test was conducted to find out if there were any significant differences in terms of Cr removal efficiency due to surface modification between treatments in the batch experiments as a whole, but a two-way factorial Anova test was performed to compare the mean of Cr removal efficiency between as-received adsorbent and two or more modified adsorbents for each type of adsorbent (CSC and/or CAC), while simultaneously detecting if there was any significant interactions between type of surface modification and either of the pertinent factors such as dose, pH, agitation speed, or contact time. After performing a one-way Anova test for preliminary evaluation, further statistical analysis was also conducted using Tukey's multiple comparisons test to compare the mean of Cr removal performance of pairwise adsorbents (CSC and/or CAC). All statistical tests were performed using SPSS 11.00 Windows version with confidence interval of 95% ($p \leq 0.05$).

CONCLUSION

Throughout this study period, all the Standard Methods (1998) described above were consistently employed to determine not only Cr concentrations after adsorption treatment, but also the characteristics of the real wastewater samples collected from a local electroplating industry in Rangsit (Thailand). The next chapter will present, analyze and discuss important findings of this study.

(Reprinted in part from *Chemosphere 54(7):951-967, Babel and Kurniawan, Cr(VI) removal from synthetic wastewater using coconut shell charcoal and commercial activated carbon modified with oxidizing agents and/or chitosan,* Copyright (2004), with permission from The Elsevier Publisher)

Chapter 4

RESULTS AND DISCUSSION

ABSTRACT

This chapter presents and discusses several major findings of the laboratory studies from a series of adsorption treatments using various types of low cost adsorbents from a local coconut industry such as coconut shell charcoal (CSC), from natural materials like zeolite, as well as commercial activated carbon (CAC) for Cr(VI) removal. Their treatment performances in this study are also evaluated and compared to those of other reported studies (1977-2002).

This chapter is divided into two sections. The first section highlights the technical applicability of CSC, zeolite as well as CAC for removing target metal from real wastewater collected from a local industry situated in Rangsit.

Using real Cr wastewater in batch studies and column operations, their metal removal performance before and after surface modification with HNO_3, H_2SO_4, and chitosan respectively was evaluated and compared. This toxic pollutant is a heavy metal of special interest due to its high toxicity and low biodegradability. The ability of all the adsorbents in as-received and chemically modified forms to meet the strict requirements of environmental legislation are evaluated in terms of their capability of generating treated effluents that contain a Cr concentration of lower than 0.25 mg/L (Thai Pollution Control Department, 2003) and/or of less than 0.05 mg/L (US EPA).

In general, this chapter reports major findings from a series of adsorption treatment employed in this study and highlights the various important factors that need to be considered thoroughly when selecting the most effective and inexpensive low cost adsorbents in order to protect the aquatic environment from heavy metal contamination due to the generation of Cr-contaminated wastewater from electroplating industries.

4.1. CHARACTERISTICS OF REAL CHROMIUM WASTEWATER

Further investigation on the Cr removal by all types of adsorbents was also conducted using real wastewater, discharged from a local electroplating industry in Rangsit, Bangkok, Thailand. The characteristics of wastewater are presented in Table 4.1.

Table 4.1 suggests that the real wastewater contained not only Cr ions, but also inorganic substances and organic compounds that might be difficult to biodegrade (as reflected by the COD value). This suggests that the existence of such impurities in real electroplating wastewater would lower the Cr adsorption capacities of adsorbents due to the presence of other contaminants.

4.2. BATCH STUDIES

4.2.1. Coconut Shell Charcoal (CSC)

4.2.1.1. Effects of Dose

The dose-dependence on the Cr removal by CSC was studied by varying the amount of dose from 1.5 g/L to 24.0 g/L at ambient temperature, while keeping other parameters (pH, agitation speed, and contact time) constant. Figure 4.1 presents the Cr removal by all types of CSC as a function of dose.

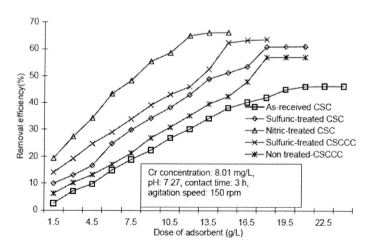

Figure 4.1. Effects of dose on the Cr removal by all types of CSC.

Table 4.1. Characteristics of wastewater
from electroplating industry

Parameter		Industrial effluent standards in Thailand*
pH	7.27	5.5-9.0
Conductivity (μs)	992	-
Turbidity (NTU)	55	< 50
Total dissolved solids (mg/L)	2,121	< 3,000 mg/L
Suspended solids (mg/L)	79.50	< 50 mg/L
COD (mg/L)	250.02	< 120 mg/L
BOD_5 (mg/L)	49.97	< 20 mg/L
Cr^{6+} (mg/L)	8.01	< 0.25 mg/L
Cr^{3+} (mg/L)	2.02	< 0.75 mg/L
Total Cr(mg/L)	10.03	< 1.00 mg/L

* Source: Thai Pollution Control Department on Industrial Effluent Standards (2003).

Figure 4.1 reveals that increasing the adsorbent dose generally resulted in an increase in the Cr removal by CSC to a certain value and then there was no increase of metal adsorption at equilibrium. At 8 mg/L of Cr concentration, the Cr removal by the nitric-treated CSC significantly improved from 19% to 65% when the dose of nitric-treated CSC increased from 1.5 g/L to 12.0 g/L. This can be attributed to the fact that surface modification of CSC with HNO_3 changed not only its textural characteristics, but also the chemical nature of its surface, due to the generation of oxygen-containing surface functional group on its surface and because of the increasing quantity of negative charge of the surface oxygen complex (Aggarwal et al., 1999).

Although it was not statistically significant difference (p>0.05; t-test), a comparison in terms of the Cr uptake of nitric-treated CSC between synthetic and real Cr wastewater at the initial Cr concentration of 10 mg/L and 8 mg/L respectively indicates that the Cr removal in synthetic wastewater was slightly higher than that of real wastewater (Figure 4.2). This difference might have been attributed to the presence of other metal ions and organic impurities in the electroplating wastewater, in which a slight hindrance to the Cr adsorption occurred, compared to the synthetic wastewater.

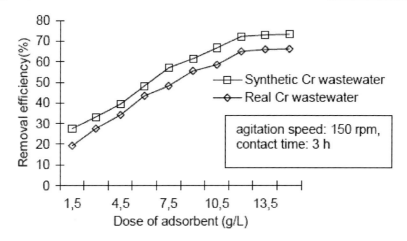

Figure 4.2. A comparison of Cr removal efficiency of the nitric-treated CSC between synthetic and real Cr wastewater based on dose.

Further investigation on the COD removal in real wastewater was conducted to characterize the effluent wastewater from electroplating industry. The COD is an accurate indication of the total oxygen consumed during the oxidation of oxidizable organic matter.

It is observed from Figure 4.3 that increasing the adsorbent concentration resulted in a lower COD value of the wastewater. The dose of nitric-treated CSC higher than 7.5 g/L was found to be effective to achieve a significant reduction of COD up to 93%. Thus, suggesting that treatment of wastewater with this type of adsorbent substantially lowered its COD value to 17.50 mg/L, which was less than 120 mg/L, the maximum effluent discharge standards of COD in Thailand.

A remarkable difference in terms of COD removal was also observed between as-received CSC and chemically modified CSC ($p \leq 0.05$, paired t-test). Of the four types of chemically modified CSC, the adsorbent chemically oxidized with nitric acid demonstrated the highest COD removal (93%) compared to others at the same Cr concentration of 8 mg/L. It is important to note that as-received CSC could still meet the effluent limit of COD, despite it has the lowest COD removal efficiency (60%). This suggests that as-received CSC could reduce the COD content of wastewater to the desired level (less than 120 mg/L), in spite of the absence of surface modifications with nitric or sulfuric acids.

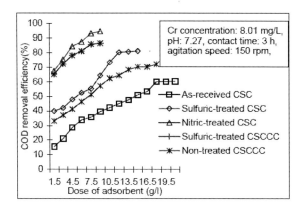

Figure 4.3. Effects of CSC dose on the COD removal efficiency.

4.2.1.2. Effects of pH

It is a well-known fact that the metal adsorption on the surface of adsorbent is dependent on pH. Figure 4.4 presents the Cr uptake by all types of CSC at different pH ranging from 2.0 to 9.0.

Figure 4.4 suggests that the Cr adsorption by CSC generally improved with an increasing pH from 2.0 to 6.0. It is noticed that the removal efficiency of nitric-oxidized CSC slightly increased from 80% to 95% over a pH ranging from 2.0 to 4.0. This could be due to the fact that an increasing quantity in the negative surface charge of CSC due to oxidative pretreatment significantly facilitated the diffusion of Cr^{3+} through columbic forces at acidic conditions. This suggests that CSC was effective for a maximum Cr removal over an acidic pH range. For this reason, it is suggested that pH determines the extent of Cr removal on the surface of CSC by providing a favorable adsorbent surface charge for the adsorption to occur.

It was also found that there was a significant reduction of Cr adsorption by the nitric-treated CSC from 66% to 21% when the pH was increased from 7.0 to 9.0. This might be due to the fact that the abundance of OH^- in the solution caused an increased hindrance to the diffusion of Cr^{3+} species onto the surface of CSC (Arulanantham et al., 1989).

A comparative study was also conducted to evaluate the effects of pH on the Cr removal by the nitric-treated CSC between synthetic and real Cr wastewater at the initial Cr concentrations of 10 mg/L and 8 mg/L respectively (Figure 4.5). It was found that there is no statistically significant difference in terms of Cr removal efficiency between the two types of Cr-contaminated water (p>0.05; t-test), in spite of the presence of other contaminants and impurities in the real wastewater.

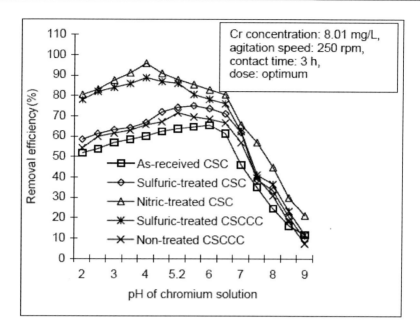

Figure 4.4. Effects of pH on the Cr removal efficiency of CSC.

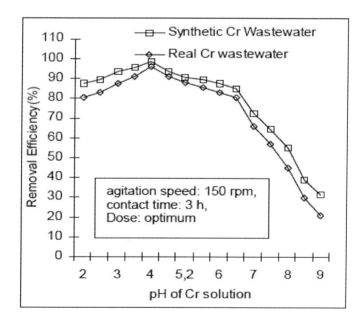

Figure 4.5. A comparison of the Cr removal efficiency of nitric-treated CSC between synthetic and real Cr wastewater based on pH.

A further study on the effects of pH on the COD removal was also carried out. Figure 4.6 shows the variation of COD removal efficiency at a different pH range from 2.0 to 9.0.

Figure 4.6 suggests that a complete COD removal by the nitric-treated CSC was attained at a pH ranging from 4.0-4.5, while its COD removal efficiency at the initial pH of wastewater (pH≈7.27) was 66%. This suggests the need for pH adjustment of the wastewater from initial pH 7.27 to a pH range of 4.0-5.0 in order to maximize COD removal, in spite of the additional operational cost due to chemicals required for pH adjustment. It was also noticed that the increase in the COD removal to less than 70% for this adsorbent was not significant at pH higher than 7.0. After treating the real wastewater with nitric-treated CSC, it may be possible to reuse the wastewater, as its COD value significantly decreased after adsorption treatment (less than 120 mg/L).

4.2.1.3. Effects of Agitation Speed

In order to study the effects of agitation speed on the Cr removal by CSC, the speed was varied from 90 rpm to 190 rpm, while keeping the optimum dose and pH as constant. Figure 4.7 illustrates the effects of agitation speed on the Cr removal by of all types of CSC.

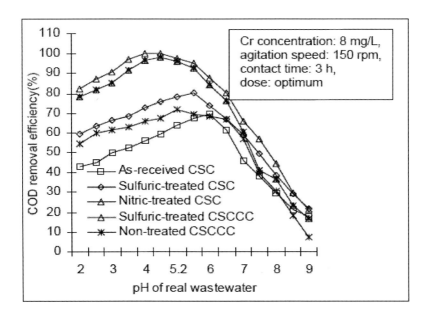

Figure 4.6. Effects of pH on the COD removal by all types of CSC.

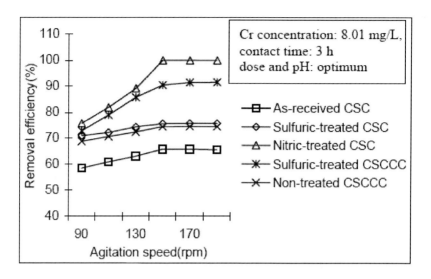

Figure 4.7. Effects of agitation speed on the Cr removal efficiency of CSC.

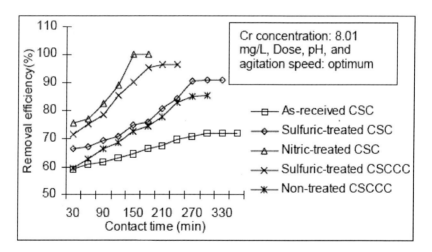

Figure 4.8. Effects of contact time on the Cr removal efficiency of CSC.

Figure 4.7 suggests that the Cr removal by the nitric-oxidized CSC significantly improved from 75% to 100%, as the speed increased from 90 rpm to 150 rpm, and then, it tended to be constant for higher agitation speeds. This can be explained due to the fact that increasing shaking speed resulted in a decrease of the boundary layer resistance to mass transfer in the bulk, causing a higher driving force for Cr(III) removal (Singh et al., 1998; Al-Degs et al,

2000). In this study, it was found that no significant difference in the Cr removal efficiency ($p > 0.05$, ANOVA test) was observed due to the degree of speed for all types of CSC.

4.2.1.4. Effects of Contact Time

Time required to attain equilibrium is important to predict the efficiency and technical feasibility of an adsorbent for metal removal. Therefore, the Cr adsorption on CSC was studied as a function of contact time.

It can be seen from Figure 4.8, that an increase in contact time from 30 min to 150 min remarkably increased the Cr removal by the nitric-treated CSC from 75% to 100%. However, a further increase in contact time had negligible effects on Cr removal by that adsorbent. This behavior suggests the occurrence of a rapid external mass transfer followed by a slower internal diffusion process (Singh et al., 1988; Meshko et al., 2001).

A significant difference of optimum contact time between CSC in chemically oxidized form and that in as-received form can be explained due to the fact that oxidative pretreatment substantially increased the number of adsorption sites of CSC and the quantity of its negative surface charge. Consequently, Cr adsorption on the carbon surface of CSC rapidly proceeded, as more Cr species were adsorbed on its surface through columbic forces at acidic condition (Peniche-Covas et al., 1992; Findon et al., 1993; Aung, 1997).

Overall, the contact time required by all types of CSC for Cr removal from real wastewater was relatively short ranging from 2 h to 6 h, suggesting a rapid transport of adsorbate species from the bulk to the surface of adsorbent.

4.2.1.5. Adsorption Isotherm Study

A. Langmuir Isotherm

The Langmuir model assumes that the uptake of metal ions occurs on a homogeneous surface by monolayer adsorption without any interaction between adsorbed ions and that the energy of adsorption proportionally corresponds to all surface sites. The Langmuir isotherm constants are calculated from the following linearized form:

$$\frac{C_e}{Q_e} = \frac{1}{a_m b} + \frac{1}{a_m} C_e \qquad (4.1)$$

where a_m and b are the Langmuir constants determined from the slope and the intercept of the plot, indicative of the maximum adsorption capacity (mg/g) of the adsorbent and the energy of adsorption respectively, while C_e is the remaining concentration of adsorbate after equilibrium (mg/L), and Q_e is the amount of metal adsorbed at equilibrium (mg/g) (Kadirvelu et al., 2001).

The single-solute sorption of Cr by CSC with varying the dose (1.5 g/L to 6 g/L) is presented in Figure 4.9.

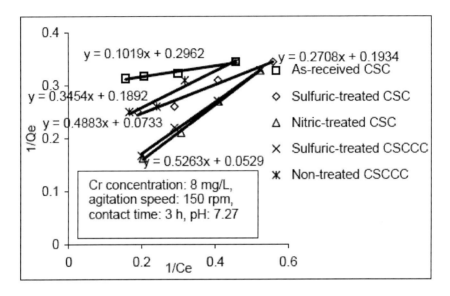

Figure 4.9. Langmuir plots for all types of CSC.

The isotherm data were analyzed using the Langmuir isotherm (Equation 4.1). The Langmuir constants (a_m and b), presented in Table 4.2, are graphically determined from the intercept and the slope of the plots. The linear plot of $1/Q_e$ against $1/C_e$ in Figure 4.9 reveals that the experimental data were representative to the Langmuir isotherm, suggesting the formation of monolayer coverage of the adsorbate particle on the adsorbent surface.

The essential characteristics of the Langmuir isotherm can be expressed in terms of a dimensionless constant separation factor (R_L) (Equation 4.2).

$$R_L = \frac{1}{1 + bC_0} \qquad (4.2)$$

where b is the Langmuir constant, while C_o is the initial concentration of metal ion. The value of R_L indicated the type of isotherm to be irreversible ($R_L=0$), favorable ($0<R_L<1$), linear ($R_L=1$), or unfavorable ($R_L>1$). The values of R_L were found to be $0<R_L<1$ for all types of CSC (Table 4.2).

B. Freundlich Isotherm

Freundlich isotherm assumes that the uptakes of metal ions occur on a heterogeneous surface by multilayer adsorption and that the amount of adsorbate adsorbed improves with an increase in the concentration of adsorbent. The Freundlich isotherm is commonly expressed as:

$$Q_e = K_f C_e^{1/n} \tag{4.3}$$

where K_f and n are the constants of the Freundlich isotherm incorporating the adsorption capacity (mg/g) and intensity, while C_e and Q_e are the remaining concentration of adsorbate after equilibrium (mg/L) and the amount adsorbed at equilibrium (mg/g) respectively. Taking logarithm from the Equation (4.3), a linearized form of the Freundlich isotherm can be represented as follows:

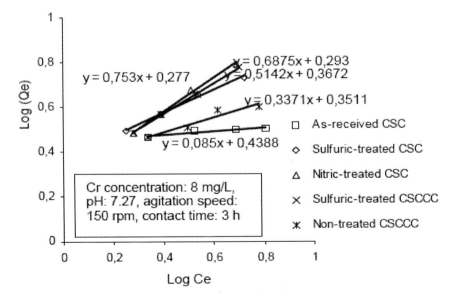

Figure 4.10. Freundlich isotherms for all types of CSC.

$$LogQ_e = \log K_f + \frac{1}{n}\log c_e \tag{4.4}$$

The straight-line plots of log Q_e versus log C_e (Figure 4.10) also indicate the applicability of the Freundlich isotherm for Cr adsorption by all types of CSC. In this study, however, the Langmuir isotherm had a better fitting model than the Freundlich, as the former had higher corresponding coefficient correlations than the latter (Table 4.2).

Kf and n, calculated from the intercept and the slope of the plots, are also presented in Table 4.2. This can be seen that the n values for all types of CSC ranged from 0.8 to 1.7. It was also pointed out that the n value of as-received CSC was less than 1, suggesting that the Cr adsorption on its carbon surface was not favorable (Kadirvelu et al., 2001).

4.2.1.6. Adsorption Mechanisms of Cr Removal by CSC

Adsorption mechanism (chemical or physical) is an important indicator to describe the type and degree of physico-chemical interactions between the adsorbate and the adsorbent in the solutions. Since the acidic functional groups of CSC, consisting of carboxylic, hydroxyl carbonyl, and lactone, present on its surface and are responsible for metal adsorption, it is assumed that the physico-chemical interactions that might occur during chromium removal by CSC could be expressed as follows:

$$M^{n+} + n\,(\text{-COOH}) \longleftrightarrow (\text{-COO})_n M + n\,H^+ \tag{4.5}$$

where (–COOH) represents the surface functional group of CSC and n is the coefficient of the reaction component depending on the oxidation state of metal ions, while M^{n+} and H^+ are the chromium and hydrogen ions respectively.

Although the oxyanion of dichromate ($Cr_2O_7^{2-}$) was employed as the source of Cr(VI) in synthetic wastewater, under common environmental pH (pH less than 6.0), the Cr(VI) exists in the predominant form of $HCrO_4^-$ (Tandon et al.,1984; Lalvani et al., 1998; Pradhan et al., 1999; Vaca-Mier et al., 2001), with the hydrolysis reaction of $Cr_2O_7^{2-}$ as follows:

$$Cr_2O_7^{2-} + H_2O \leftrightarrow 2\,HCrO_4^-\ \ pK_3 = 14.56 \tag{4.6}$$

Table 4.2. Chromium adsorption capacities (mg/g) of coconut shell charcoal (CSC) for real wastewater

Type of CSC	Langmuir isotherm						Freundlich isotherm			
	Q_e (mg/g)	R^2	a_m	b	R_L	Equation of Q_e	Q_e (mg/g)	R^2	n	Equation of Q_e
As-received	3.22*	0.9967	3.2209	2.9064	0.0412	$Q_e = \dfrac{9.3612 C_e}{(1+2.9064 C_e)}$	3.03**	0.9845	0.7695	$Q_e = 3.0311\, C_e^{0.7695}$
Coated with chitosan	5.02*	0.9876	5.0197	0.5476	0.1858	$Q_e = \dfrac{2.7488 C_e}{(1+0.5476 C_e)}$	3.45**	0.9632	0.9664	$Q_e = 3.4527\, C_e^{0.9664}$
Oxidized with sulfuric acid	5.18*	0.9932	5.1812	0.3051	0.2906	$Q_e = \dfrac{1.5807 C_e}{(1+0.3051 C_e)}$	3.76**	0.97217	1.2447	$Q_e = 3.7612\, C_e^{1.2447}$
Oxidized with sulfuric and coated with chitosan	8.32*	0.9764	8.3214	0.1502	0.4542	$Q_e = \dfrac{1.2499 C_e}{(1+0.1502 C_e)}$	4.23**	0.96421	1.4546	$Q_e = 4.2309\, C_e^{1.4546}$
Oxidized with nitric acid	13.69*	0.9854	13.6896	0.1006	0.5541	$Q_e = \dfrac{1.3772 C_e}{(1+0.1006 C_e)}$	4.97**	0.9642	1.7281	$Q_e = 4.9711\, C_e^{1.7281}$

* significant difference ($p \leq 0.05$; paired t-test) between as-received and chemically modified CSC for the Langmuir isotherm.

** no difference ($p > 0.05$; paired t-test) between as-received CSC and chemically modified CSC for the Freundlich isotherm.

In acidic solutions where intimate physico-chemical contact between the adsorbent and the adsorbate occurred, Cr(VI) demonstrated a very high positive redox potential (E^o) in the range of 1.33 V and 1.38 V (Kotaś and Stasicka, 2000), thus implying that Cr(VI) is strongly oxidizing and unstable in the presence of electron donors (Rai et al., 1989). The carbon surface of CSC contains carboxylic and hydroxyl groups, which played roles as electron donors in the solution (Tan et al., 1993). Consequently, Cr(VI) oxyanion is readily reduced to Cr(III) ions due to the presence of electron donors of CSC according to the following reaction of electron transfer:

$$\text{pH 5.0-6.0}$$
$$HCrO_4^- + 7\,H^+ + 3e^- \leftrightarrow Cr^{3+} + 4\,H_2O \quad E^o = 1.20 \text{ V} \qquad (4.7)$$

This can be seen from Equation (4.7) that the reduction of Cr(VI) oxyanion is accompanied by a large amount of proton consumption in the acidic solutions, thus confirming the decisive role played by H^+ in the Cr(VI) removal by CSC. The latter mechanism was also reported by Ik and Zoltek (1977), Huang (1977), Archundia et al. (1993), Ramos et al. (1994), Sharma and Forster (1994b; 1996), and Reddad et al. (2003).

It was also reported that the redox potential (E^o) of the Cr(VI)/Cr(III) strongly depends on pH; at pH≈1, E^o≈1.3 V and at pH≈5, E^o≈0.68 V (Lakatos et al., 2002), indicating that improving the redox potential of the oxidant would extend the oxidation towards the more resistant surface functionalities.

If the system does not have a suitable buffer capacity, the pH will continuously increase with a concomitant decrease in the redox potential of the Cr(VI)/Cr(III) system until the reaction completely stops. Lakatos et al. (2002) reported that the 0.1 M acetic acid-sodium acetate buffer at pH 5 employed in this study could supply proton for the redox reactions. However, the low concentration of protons (10^{-5}M) resulted in a lower redox potential. Thus, suggesting that the extents to which Cr reduction and oxidation were controlled by the pH.

Since the pH of synthetic wastewater ranges from 5.0 to 6.0, the Cr(III) mostly exists as $[Cr(OH)]^{+2}$ species (Cotton and Wilkison, 1988; Dean and Tobin, 1999; Kocaoba and Akcin, 2002; Lakatos et al., 2002) with the hydrolysis reaction of Cr(III) as follows:

$$Cr^{3+} + H_2O \leftrightarrow [Cr(OH)]^{2+} + H^+ \quad pK_1 = 3.85 \qquad (4.8)$$

The second mechanism controlling the adsorption of Cr(III) on the carbon surface of CSC is represented as:

$$Cr^{3+} + H_2O \longleftrightarrow [Cr(OH)]^{+2} + H^+ \tag{4.9}$$

$$HA + [Cr(OH)]^{+2} \longleftrightarrow ([Cr(OH)]^{+2}\text{-}A^-) + H^+ \tag{4.10}$$

where A represents an adsorption site on the acidic surface of CSC.

Combination of the Equations (4.9 and 4.10) gives an overall reaction as follows:

$$Cr^{3+} + H_2O + HA \longleftrightarrow ([Cr(OH)]^{+2}\text{-}A^-) + 2\ H^+ \tag{4.11}$$

The Equation (4.11) suggests that the Cr adsorption on the surface of CSC was due to columbic forces between the positive charge of Cr(III) and the negative surface charge of CSC. Based on the attractive electrostatic interactions between the electron-donating nature of the oxygen-containing functional groups on the surface of CSC acting as a Lewis base and the electron-accepting nature of heavy metal ions (Lewis acid), the ion exchange mechanism of Cr might be preferentially considered (Nakano et al., 2001). For instance, a trivalent chromium ion might attach itself to three adjacent hydroxyl groups, carboxylic groups, and oxyl groups, which can donate lone pairs of delocalized π electrons on the carbon surface to the metal ion for the formation of surface oxide compounds ($[Cr(OH)]^{+2}\text{-}A^-$) (Shukla et al., 2002; Reddad et al., 2003).

Equation (4.11) indicates that as the solution pH increased to 6.0 or the concentration of hydrogen ions decreased (to 10^{-6}M), the adsorption reaction shifted from left to right, which resulted in the production of more oxygenated Cr complex ($[Cr(OH)]^{+2}\text{-}A^-$) on the surface of CSC or higher Cr removal by the adsorbent. After equilibrium, it was found that the pH of the Cr solution decreased after adsorption treatment. This indicates that the CSC was hydrophilic and acidic in nature, as more H^+ was released into the solution.

4.2.1.7. Effects of Surface Oxidation on the Sorptive Capacity of CSC

The carbon surface of CSC has unsaturated C=C bonds, which on oxidation can generate more oxygen-containing surface functional groups and increase the surface area available for Cr uptake by improving its pore structure (Nakano et al., 2001). The increase in the concentration of potentially accessible carboxyl sites due to surface oxidation was clearly demonstrated by the increase of Cr adsorption capacity in the isotherm studies.

As the Cr(VI) reduction mostly took place with the carbon bound to –OH surface functionalities (Lakatos et al., 2002), the oxidation of CSC also resulted in an increase in the concentration of electron donor site. Due to the presence of delocalized π-electrons that is easily transferable in a conjugated system of aromatic bonds, CSC has a remarkable electrical conductivity, as localization of the $2p$ electrons of the functional group-oxygen (-COO$^-$) into the π-conjugated system results in an increase in the negative charge present on the surface. Therefore, a competition occurred between H$^+$ and Cr(III) ions for the adsorption sites (El-Hendawy, 2003). Consequently, the availability of relatively high amount of -COO$^-$ renders the adsorbent more acidic and have more negative surface charge; thus, increasing the sorptive capacity of the oxidized CSC on Cr(III) ions through columbic interactions.

4.2.2. Zeolite

4.2.2.1. Effects of Dose

To investigate the Cr adsorption capacity of zeolite, real wastewater was also mixed with zeolite with varying the dose (from 1.5 g/L to 24 g/L). The Cr removal by both types of zeolite were evaluated and statistically compared.

The results, presented in Figure 4.11, show that the Cr removal by as-received zeolite slightly increased from 2% to 36% with an increase in the amount of dose from 1.5 g/L to 21 g/L. Beyond this dose, the Cr removal by as-received zeolite tended to be constant. Thus, suggesting that equilibrium conditions had been reached. Similarly, the Cr removal by the NaCl-treated zeolite increased from 7% to 46% with increasing the dose from 1.5 g/L to 18 g/L. Therefore, 21 g/L and 18 g/L of as-received zeolite and NaCl-treated zeolite respectively were selected for the remaining studies.

A comparative study is also conducted to evaluate the Cr removal by the NaCl-treated zeolite between synthetic and real Cr wastewater at the initial Cr concentration of 10 mg/L and 8 mg/L respectively (Figure 4.12). Statistical analysis reveals that there was no statistically significant difference ($p > 0.05$; independent t-test) between the two types of Cr-contaminated water, despite the facts that the wastewater contained other metal contaminants and impurities, which might hinder the Cr removal by zeolite (Table 4.3) and that the initial pH of real wastewater (pH$\approx$7.27) and synthetic one (pH$\approx$5.20) was different from each other.

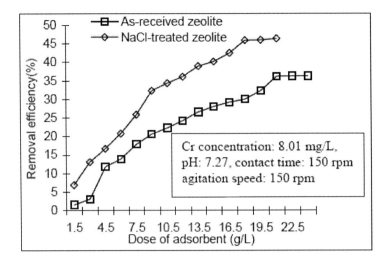

Figure 4.11. Effects of dose on the Cr removal efficiency of zeolite.

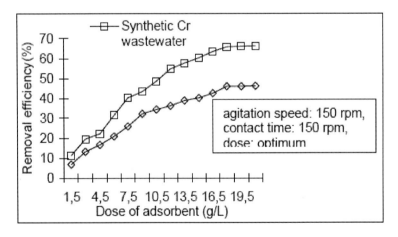

Figure 4.12. A comparison of Cr removal efficiency of NaCl-treated zeolite between synthetic and real Cr wastewater based on dose.

A further study investigating the dose-dependence on the COD removal by zeolite was also carried out. The results are presented in Figure 4.13.

It is apparent from Figure 4.13 that increasing the zeolite concentration resulted in a lower COD value. A remarkable difference of COD removal efficiency was also observed for both types of zeolite ($p \leq 0.05$; paired t-test). The dose of NaCl-treated zeolite higher than 15 g/L was found to slightly reduce the COD up to 40% at COD concentration of 250 mg/L. Thus suggesting that treatment of real wastewater with the NaCl-treated zeolite was not effective enough to significantly reduce the COD of wastewater, as the final COD value (150 mg/L) was still higher than 120 mg/L, the maximum effluent discharge standards of COD in Thailand.

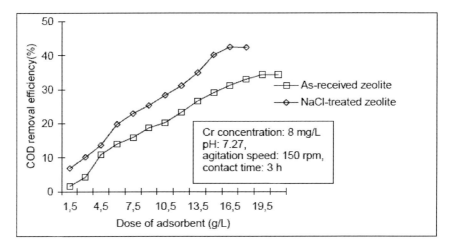

Figure 4.13. Effects of zeolite dose on the COD removal efficiency.

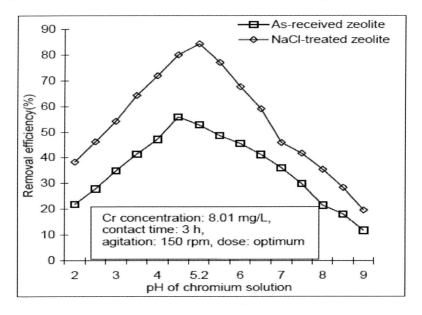

Figure 4.14. Effects of pH on the Cr removal efficiency of zeolite.

4.2.2.2. Effects of pH

Knowledge of the optimum pH is also important due to the dependence of the type of hydrolysis species on pH. The effects of pH on Cr removal by zeolite are presented in Figure 4.14.

It is evident from the above figure that the Cr removal by zeolite was also significantly affected by the change of pH, as the Cr removal by the NaCl-treated zeolite remarkably increased from 39% to 84% with increasing the pH from 2.0 to 5.2; suggesting that the Cr removal by the NaCl-treated zeolite was very effective at pH lower than 7.0. However, it substantially decreased to 20% when the pH was increased beyond 7.0.

This can be due to the fact that at acidic conditions, the zeolite surface is highly protonated, which is a favorable condition for Cr removal. Consequently, the negatively charged surface of zeolite started acquiring a net positive charge, making the situation electrostatically favorable for a higher Cr uptake. However, at pH greater than 7.00, the zeolite surface became less negatively charged due to deprotonation. This reduced the columbic forces between the negative charge of the zeolite and Cr^{3+}. As a result, Cr adsorption on the zeolite surface was hindered and resulted in a lower Cr uptake by the zeolite (Misaelides et al., 1995).

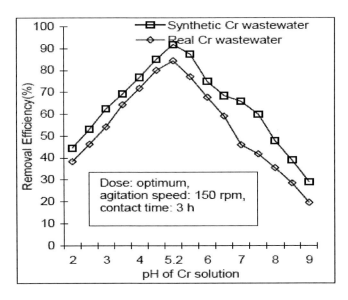

Figure 4.15. A comparison of Cr removal efficiency of NaCl-treated zeolite between synthetic and real Cr wastewater based on pH.

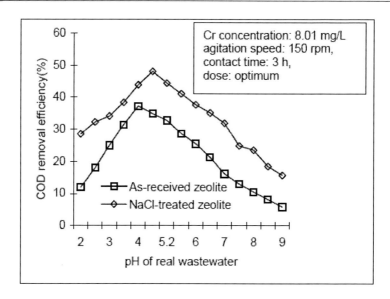

Figure 4.16. Effects of pH on the COD removal by both types of zeolite.

Although statistical analysis confirms that there was no significant difference (p>0.05; t-test), a comparison in terms of the Cr uptake by the NaCl-treated zeolite between synthetic and real Cr wastewater at the initial Cr concentrations of 10 mg/L and 8 mg/L respectively indicates that the Cr removal in synthetic wastewater was slightly higher than in the real wastewater (Figure 4.15). This might be attributed to the presence of few impurities and trace metal contaminants in the real wastewater, enabling most chromium ions to be adsorbed on the surface of zeolite without any interfering effects. In addition, the difference in terms of the initial pH between real wastewater (pH≈7.27) and synthetic one (pH≈5.20) might also contribute to the different Cr removal between the two types of Cr-contaminated wastewater.

Further investigation on the effects of pH on the COD removal in real wastewater was also conducted. Figure 4.16 shows the variation of COD removal efficiency at varying pH from 2.0 to 9.0.

Figure 4.16 suggests that the highest COD removal by the NaCl-treated zeolite (48%) was attained at pH 4.5. However, the increase in COD removal was not significant at pH higher than 7.0. After treating the real wastewater with this adsorbent, the quality of the wastewater was found to be insufficient for possible reuse, as the COD of the treated effluent (130 mg/L) was still higher than the maximum effluent limit of COD in Thailand.

4.2.2.3. Effects of Agitation Speed

The variations of Cr uptake by zeolite with a different agitation speeds from 90 rpm to 190 rpm are presented in Figure 4.17.

It was observed from Figure 4.17 that the Cr uptake by the NaCl-treated zeolite slightly improved from 51% to 84% with increasing the speed from 90 rpm to 150 rpm. This could be due to the fact that increasing agitation speed reduced the film boundary layer surrounding the adsorbent's particle. As a result, the external film mass transfer coefficient remarkably increased and resulted in a substantial increase in the rate of Cr removal by the zeolite (Singh et al., 1988).

4.2.2.4. Effects of Contact Time

Knowledge of the rate of Cr uptake to attain equilibrium condition is essentially necessary to characterize an adsorbent. Therefore, the removal of Cr was studied as a function of contact time. This study was conducted at optimum condition of dose, pH and speed where a maximum Cr removal by zeolite could take place.

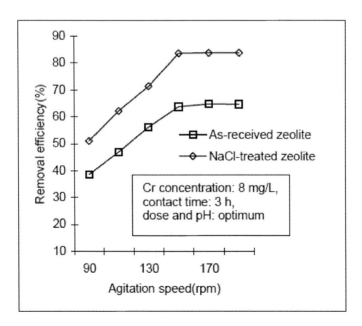

Figure 4.17. Effects of agitation speed on the Cr removal efficiency of zeolite.

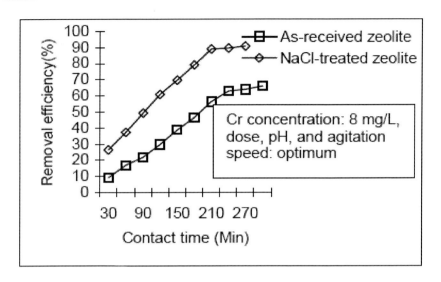

Figure 4.18. Time-dependent of Cr adsorption on zeolite.

Figure 4.18 suggests that the equilibrium for the NaCl-treated zeolite was achieved within 210 min. Thus indicates that the rate of Cr uptake was rapid in the early stage and gradually decreased and then became constant at equilibrium. It is important to note that adsorption equilibrium is attained when the concentration of adsorbate in the bulk solution is in dynamic balance with that at the interface.

As depicted in Figure 4.18, the equilibrium between Cr and zeolite was attained within 240 min, where the NaCl-treated zeolite had a higher Cr removal efficiency (90%) than as-received zeolite (63%). No significant increase of Cr removal occurred after 240 min of contact time ($p > 0.05$; paired t test).

4.2.2.5. Adsorption Isotherm Study

A. Langmuir Isotherm

Figure 4.19 illustrates the isotherm study of Cr adsorption by zeolite with varying the dose (from 0.5 g/L to 2 g/L). All the Langmuir constants, presented in Table 4.3, are determined from the intercept and the slope of the linear plots of $1/Qe$ and $1/Ce$.

Table 4.3. Chromium adsorption capacities (mg/g) of zeolite for real wastewater

Type of zeolite	Langmuir isotherm						Freundlich isotherm			
	R^2	a_m	b	R_L	Equation of Q_e	Q_e (mg/g)	R^2	n	Equation of Q_e	Q_e (mg/g)
As-received	0.9979	3.0612	4.5532	0.1091	$Q_e = \dfrac{3.1261 C_e}{(1+4.5532 C_e)}$	3.06*	0.9754	1.9271	$Q_e = \dfrac{2.6714}{C_e^{1.9271}}$	2.67*
Treated with NaCl	0.9993	5.9691	0.8041	0.2593	$Q_e = \dfrac{2.1316 C_e}{(1+0.8041 C_e)}$	5.24*	0.9886	3.9343	$Q_e = \dfrac{5.2397}{C_e^{3.9343}}$	5.97*

*significant difference ($p \le 0.05$; paired t-test) between as-received zeolite and NaCl-treated zeolite for both the Langmuir and Freundlich isotherms.

Table 4.4. Chromium adsorption capacities (mg/g) of commercial activated carbon (CAC) for real wastewater

Type of CAC	Langmuir isotherm						Freundlich isotherm			
	R^2	a_m	b	R_L	Equation of Q_e	Q_e (mg/g)	R^2	n	Equation of Q_e	Q_e (mg/g)
As-received	0.9975	6.0397	1.0212	0.1091	$Q_e = \dfrac{6.1677 C_e}{(1+1.0212 C_e)}$	6.04*	0.9774	3.8897	$Q_e = \dfrac{4.2411}{C_e^{3.8897}}$	4.24*
Oxidized with sulfuric acid	0.9921	9.0212	0.3571	0.2593	$Q_e = \dfrac{3.2215 C_e}{(1+0.2357 C_e)}$	9.02*	0.9671	1.7493	$Q_e = \dfrac{7.1698}{C_e^{1.7943}}$	7.17*
Oxidized with nitric acid	0.9789	14.0296	0.0871	0.5893	$Q_e = \dfrac{1.2220 C_e}{(1+0.0871 C_e)}$	14.03*	0.9953	1.2772	$Q_e = \dfrac{14.9641}{C_e^{1.2772}}$	14.96*

*significant difference ($p \le 0.05$; paired t-test) between as-received and chemically modified CAC for both the Langmuir and Freundlich isotherms.

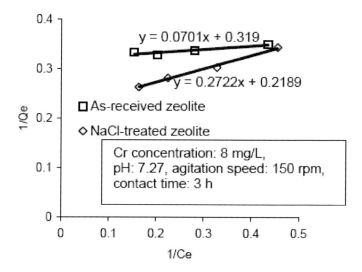

Figure 4.19. Langmuir plots of Cr adsorption on zeolite.

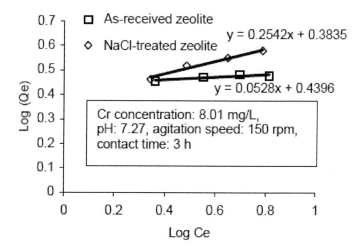

Figure 4.20. Freundlich isotherm for all types of zeolite.

When the experimental data were plotted against the Langmuir isotherm, they were found to be applicable. Conformity of the data to the Langmuir isotherm indicates that Cr adsorption by zeolite could be characterized as monolayer.

Figure 4.19 also indicates that the NaCl-treated zeolite has a better fitting of the Langmuir model than the as-received zeolite. This suggests that the

Langmuir isotherm was more representative for the treated adsorbent than the latter. The constant values of R_L in the Langmuir isotherm suggest that a favorable Cr adsorption ($0<R_L<1$) took place on the zeolite surface (Table 4.3).

B. Freundlich Isotherm

The Cr removal by zeolite can be also mathematically expresses by plotting the data into a straight-line based on the Freundlich isotherm equation (Figure 4.20). The constant values (n) of the Freundlich isotherm, lying between 1 and 10, suggest that the Cr adsorption on the zeolite surface was also favorable (Table 4.3).

Between the two isotherms, the Langmuir isotherm was more representative for zeolite than the Freundlich. This might be attributed to the fact that zeolite has a smaller surface area for Cr adsorption (780 m^2/g). Consequently, only monolayer adsorption occurred on its surface, regardless of any chemical pretreatment.

4.2.2.6. Effects of Chemical Treatment

Zeolite contains a complement of exchangeable sodium, potassium, magnesium, and calcium ions, with the selectivity of metals as follow: $K^+>Mg^{2+}>Ca^{2+}>Na^+$. To prepare the zeolite in the homoionic form of Na^+, zeolite was treated with NaCl before adsorption. This treatment was conducted based on the findings of previous studies that Na^+ was the most effective exchangeable ion for facilitating ion exchange with target heavy metal (Blanchard et al., 1984; Zamzow et al., 1990; Mondales et al., 1995).

Based on the literature survey, it was reported that after pretreatment with NaCl, certain cations such as K^+ and Ca^{2+} were strongly held by the zeolite in preference to Na^+. Therefore, Na^+ was mostly involved in the ion exchange process. It was pointed out that the exposure of zeolite to highly concentrated-NaCl led to the production of the sodium-rich samples and that phenomenon led to the exchange of calcium ions. It was further mentioned that Mg^{2+} and K^+ were not exchangeable with other cations, as they were associated with impurities in the samples. For this reasons, K^+, Ca^{2+}, and Mg^{2+} contents in the zeolite could be ignored (Semmens and Martin, 1988).

4.2.2.7. Steps of Chromium Removal by Zeolite

There are two major consecutive steps of Cr removal by the zeolite. The first step is as discussed in the Equations (4.7 and 4.8), while the second step controlling the Cr removal by zeolite is represented as follows:

$$[Cr(OH)]^{+2}{}_{(s)} + Na_nA_{(z)} + n\, H_2O_{(s)} \leftrightarrow$$

$$([Cr(OH)]^{+2}\text{-}H_n\text{-}A)_{(z)} + n\, Na^{+}{}_{(s)} + n\, OH^{-}{}_{(s)} \qquad (4.12)$$

where A and n represent the adsorption sites on the zeolite surface and the coefficients of reaction component respectively, while subscripts s and z denote "solution" and "zeolite" phases respectively. The Equation (4.12) suggests that the negative charge of the zeolite, which resulting from the tetrahedrally-coordinated alumunium, has been balanced by the exchangeable Cr^{3+}. Thus suggesting that the Cr uptake by zeolite mostly occurred due to ion exchange, although some Cr removal by zeolite might be due to adsorption also.

As pH increased to 4.5, the adsorption reaction shifted from left to right, which resulted in the production of more surface complex ($[Cr(OH)]^{+2}\text{-}H_n\text{-}A$) on zeolite. It was also found that the final pH of solution slightly increased after adsorption reaction. It could be due to the fact that the hydrolysis reaction of zeolite caused more OH^- release into the solutions, resulting in a higher Cr removal by the adsorbent. The presence of OH^- in the solution was found to make Cr^{3+} to be readily accommodated in the surface lattice of zeolite. This indicates that the removal of chromium by zeolite is pH dependent.

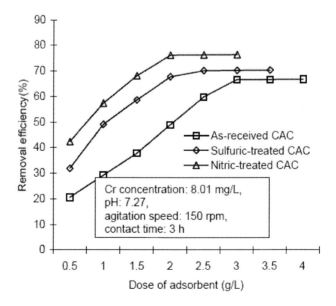

Figure 4.21. Effects of CAC dose on the removal efficiency of Cr.

4.2.3. Commercial Activated Carbon (CAC)

4.2.3.1. Effects of Dose

To study the dependence of Cr removal on dose, real wastewater was also mixed with varying the dose of CAC (0.5 to 3.5 g/L) for 3 h, while keeping other parameters as constant. The results are presented in Figure 4.21.

It is apparent from Figure 4.21 that about 76% of Cr removal was achieved within 3 h of contact time when the dose of nitric-treated CAC was increased to 2.0 g/L. For achieving maximum Cr removal, this dose was significantly lower compared to the nitric-treated CSC (12.0 g/L) and the NaCl-treated zeolite (18.0 g/L).

This could be attributed to the fact that surface modification of CAC with nitric acid has significantly improved its removal capabilities for Cr, as the oxidizing agent was capable of increasing the quantity of the negative surface charge of CAC substantially and generating more surface oxide complex on its surface (Strelko et al., 2002). Consequently, CAC substantially acquired more negative charge on its carbon surface for columbic forces with Cr^{3+} at acidic conditions, resulting in a higher Cr uptake.

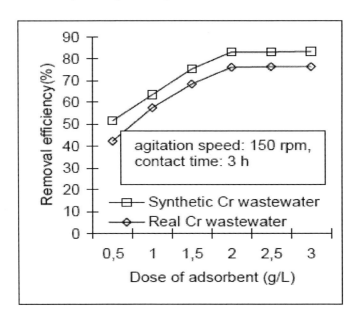

Figure 4.22. A comparison of Cr removal efficiency of nitric-treated CAC between synthetic and real Cr wastewater based on dose.

Statistical analysis also confirms that a remarkable difference in terms of Cr removal efficiency was observed among the three types of CAC ($p \leq 0.05$; ANOVA test). Of the two types of chemically modified CAC, CAC oxidized with nitric acid has greater Cr removal efficiency compared to that oxidized with sulfuric acid.

Further investigation was also carried out to evaluate and compare the Cr removal by the nitric-treated CAC between synthetic and real wastewater at the initial Cr concentration of 10 mg/L and 8 mg/L (Figure 4.22). It was found that there was no significant difference in terms of Cr removal efficiency between the two types of Cr-contaminated water ($p > 0.05$; t-test), in spite of the presence of other metal contaminants and impurities in the real wastewater.

A further study on the COD removal by all types of CAC was also performed and the results are presented in Figure 4.23.

It can be observed from the above figure that the COD removal by nitric-treated CAC slightly increased from 77% to 100% with increasing the dose from 0.5 g/L to 2 g/L, thus suggesting that the dose of the nitric-treated CAC higher than 2 g/L was effective enough to substantially lower the COD of treated effluent.

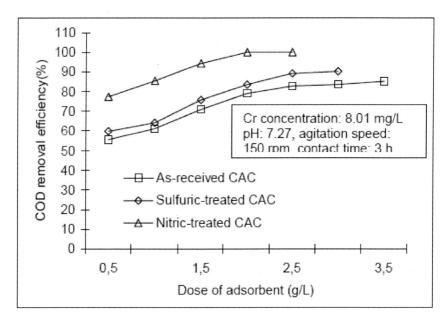

Figure 4.23. Effects of CAC dose on the COD removal efficiency.

4.2.3.2. Effects of pH

The pH of solution is also an important variable, which controls metal adsorption at the adsorbent-water interface. Therefore, the role of H^+ concentration was examined at a pH ranging from 2.0 to 9.0.

The results, presented in Figure 4.24, reveal that Cr adsorption on the surface of CAC was dependent on pH. The Cr removal by the nitric-treated CAC slightly improved from 86% to 99% when the pH increased from 2.0 to 4.0. Beyond pH 7.0, there was a remarkable reduction of Cr removal up to 36% at an initial concentration of 8 mg/L. This could be due to the fact that at acidic pH, a significantly high electrostatics attractive interactions existed between the negative surface charge of CAC and Cr^{3+}, causing more formation of the oxide surface complex on the CAC surface. As the surface charge of CAC became more negative with an increasing pH, more Cr^{3+} was substantially attracted for adsorption. This suggests that columbic forces and surface complexation played major roles in removing Cr from the solutions.

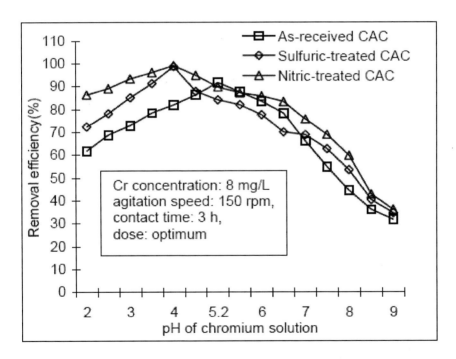

Figure 4.24. Effects of pH on the Cr removal by CAC.

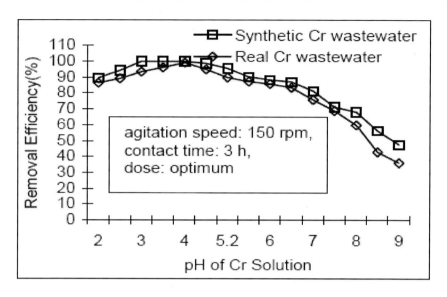

Figure 4.25. A comparison of Cr removal efficiency of nitric-treated CAC between synthetic and real Cr wastewater based on pH.

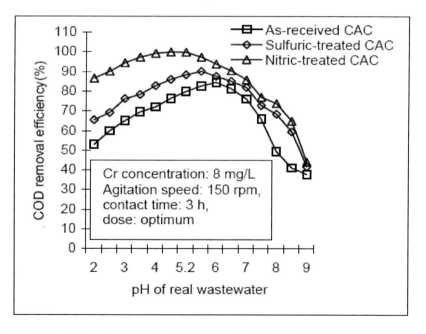

Figure 4.26. pH dependence on the COD removal efficiency of CAC.

A comparison of Cr removal by the nitric-treated CAC between synthetic and real wastewater at the initial Cr concentrations of 10 mg/L and 8 mg/L respectively was also evaluated (Figure 4.25). Although there was no statistically significant difference of Cr removal between the two types of Cr-contaminated water, it was found that the Cr removal by the nitric treated CAC followed other trends with a slightly higher Cr removal for synthetic wastewater.

Effects of pH on the COD removal by the nitric-treated CAC were also studied. The findings are presented in Figure 4.26.

It is observed from Figure 4.26 that a complete COD removal by the nitric-treated CAC was attained at pH 4.5. Beyond pH 4.5, the COD removal tended to decrease. Due to the improved quality after adsorption treatment (as indicated by the lower COD value), the treated effluent could be reused for other industrial process.

4.2.3.3. Effects of Agitation Speed

The effects of agitation speed on metal adsorption were investigated by varying the speed from 90 rpm to 190 rpm. The results are presented in Figure 4.27.

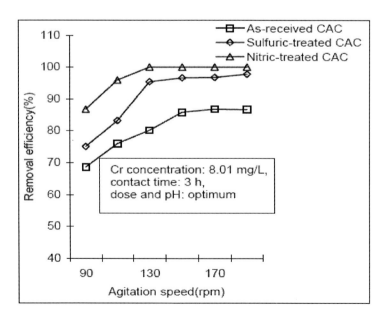

Figure 4.27. Effects of agitation speed on the Cr removal efficiency of CAC.

Figure 4.27 suggests that the Cr removal by the nitric-treated CAC was slightly enhanced from 86% to 100% when the speed increased from 90 rpm to 130 rpm. It was also observed that chemically oxidized CAC with nitric and/or sulfuric acid could achieve a maximum Cr removal at a lower speed (130 rpm) than as-received CAC (150 rpm). This could be due to the fact that oxidative pretreatment with these oxidizing agents, enabled CAC to have a rapid Cr adsorption on its carbon surface due to the formation of more adsorption sites and because of the increasing negative surface charge on the surface.

Although agitation speed affected the degree of physico-chemical interactions between the adsorbate and the adsorbent in the solution, statistical analysis confirms that there was no significant difference in terms of Cr removal among the five agitation speeds for all types of CAC ($p > 0.05$; ANOVA test).

4.2.3.4. Effects of Contact Time

Adsorption is a slow process and adequate contact time is necessary to allow the system to attain an equilibrium. Therefore, Cr adsorption by CAC was studied by varying the equilibrium time from 30 min to 180 min.

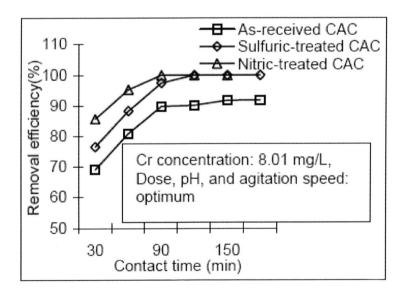

Figure 4.28. Dependence of Cr sorption on contact time.

It is apparent from Figure 4.28, an almost complete Cr removal could be attained by the nitric-treated CAC within 90 min and then its Cr removal tended to be constant for a longer contact time. This suggests that this equilibrium time (90 min) was significantly faster than that of as-received CAC (180 min). This could be due to the fact that oxidative pretreatment of CAC with nitric acid increased the number of adsorption sites on CAC (Goyal et al., 2001). Consequently, adsorption on the surface of the nitric-treated CAC proceeded more rapidly than that of the as-received CAC.

4.2.3.5. Adsorption Isotherm Study

A. Langmuir Isotherm
The linear plots of $1/Ce$ against $1/Qe$ values for all types of CAC at varying dose (from 0.5 g/L to 2.0 g/L) are presented in Figure 4.29.

In spite of their large surface area, the Langmuir isotherm was found to be representative over the equilibrium data of the adsorbents, as reflected by the correlation coefficients (Table 4.4). This table suggests that the Cr sorption by all types of CAC in the present study was favorable $(0<R_L<1)$.

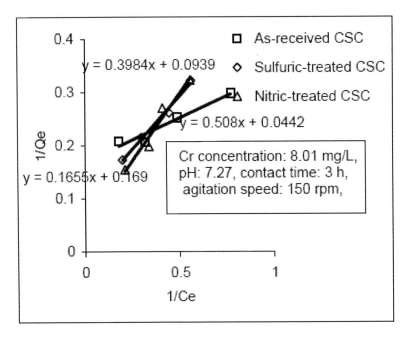

Figure 4.29. Langmuir plots for chromium adsorption on CAC.

B. Freundlich Isotherm

A straight-line plot of log Ce against log Qe also shows that the Freundlich isotherm was also applicable to the Cr adsorption by CAC (Figure 4.30).

The n values for all types of CAC, presented in Table 4.4, were in the range of 1 to 4; thus, indicating the suitability of the system under investigation. It was also found that the Freundlich isotherm was more representative for the nitric-treated CAC than the Langmuir, due to the fact that nitric is a stronger oxidizing agent than sulfuric acid; thus, enabling it to facilitate the formation of more surface site on the surface for Cr adsorption.

4.2.3.6. Adsorption Mechanism of CAC on Cr

Since CAC is made up of coconut shell and has the same types of surface functional groups as CSC, the adsorption mechanism of CAC on chromium is suggested to be the same as that of CSC (Section 4.2.1.6).

4.2.3.7. Comparison of the Cr Adsorption Capacity and COD Removal Efficiency of CSC, Zeolite, and CAC in Batch Studies for Real Wastewater

It is evident from the batch studies and Table 4.5 that the Cr adsorption capacities of all adsorbents in real wastewater with the initial chromium concentration of 8 mg/L are fairly comparable to that in synthetic wastewater with 10 mg/L of metal concentration. Statistical analysis reveals that there was no significant difference in terms of Cr removal for all the investigated adsorbents ($p > 0.05$; independent t-test) between the two types of Cr-contaminated water; thus suggesting that the presence of trace metal contaminants and other impurities in the real wastewater did not pose interfering effects on the Cr removal by all adsorbents.

Table 4.5 suggests that both chemically modified CSC with sulfuric acid and coated with chitosan (sulfuric-treated CSCCC) and/or nitric acid have significantly improved Cr removal capabilities of 8.32 mg/g and 13.69 mg/g, which were comparable to that of sulfuric-treated CAC and nitric-treated CAC (9.02 mg/g and 14.96 mg/g) respectively.

Among the four types of chemically modified CSC, the nitric-treated CSC and the sulfuric-treated CSC coated with chitosan (sulfuric-treated CSCCC) were also capable of significantly lowering the initial COD of real wastewater (250 mg/L) up to 95% and 86 %, respectively (Figure 4.31). Thus enabling the treated wastewater for reuse, as the final COD value was less than 120 mg/L, the maximum effluent discharge standard of COD in Thailand.

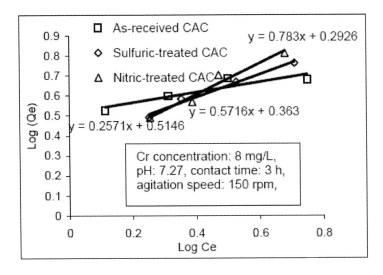

Figure 4.30. Freundlich isotherm for all types of CAC.

Table 4.5. Chromium adsorption capacities of CSC, zeolite, and CAC for real wastewater

Type of adsorbent	Type of surface modification	Real chromium wastewater (8 mg/L)			
		Langmuir isotherm		Freundlich isotherm	
		Q_e(mg/g)	R^2	Q_e(mg/g)	R^2
CSC	As-received	3.22*	0.9967	3.03**	0.9845
	Coated with chitosan	5.02*	0.9876	3.45**	0.9632
	Oxidized with H_2SO_4	5.18*	0.9932	3.76**	0.9721
	Oxidized with sulfuric acid and coated with chitosan	8.32*	0.9764	4.23**	0.9642
	Oxidized with HNO_3	13.69*	0.9854	4.97**	0.9642
Zeolite	As-received	3.06*	0.9979	2.77*	0.9754
	Treatment with NaCl	5.97*	0.9993	5.04*	0.9886
CAC	As-received	6.04*	0.9975	4.24*	0.9774
	Oxidized with H_2SO_4	9.02*	0.9921	7.17*	0.9671
	Oxidized with HNO_3	14.03*	0.9789	14.96*	0.9953

Remarks: *Significant difference ($p \leq 0.05$; paired t-test) between CSC and CAC in as-received forms and their chemically modified forms for the Langmuir and the Freundlich isotherms

* No difference ($p > 0.05$; paired t-test) between as-received CSC and zeolite and its chemically modified forms for Freundlich isotherm.

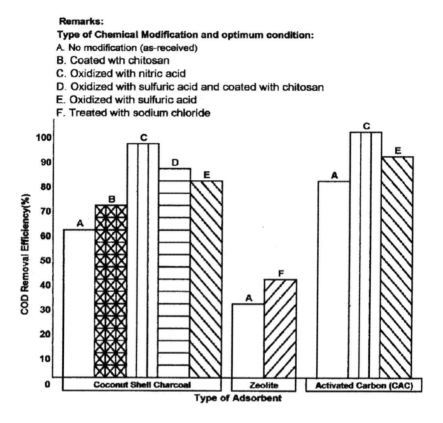

Figure 4.31. Summary of the COD removal efficiency (mg/g) for all adsorbents.

Zeolite in as-received form had a reasonable chromium removal for real wastewater, as it has a comparable Cr adsorption capacity (3.06 mg/g) to as-received CSC (3.22 mg/g) (p>0.05; t-test). However, the zeolite has poor performances for COD removal. The optimum dose of the NaCl-treated zeolite (15.0 g/L) was found to slightly reduce the COD in real wastewater up to 40% of the initial COD of 250 mg/L, which was still higher than the maximum effluent discharge standard of COD (less than 120 mg/L) in Thailand. This suggests the ineffectiveness of treating the real wastewater with zeolite.

Although CAC in as-received form demonstrated an outstanding Cr adsorption capacity (6.04 mg/g), surface modification of CAC with a strong oxidizing agent could still remarkably improve its Cr adsorption capacity. It was found that CAC oxidized with nitric (14.96 mg/g) has one and half times higher adsorption capacity than that oxidized with sulfuric acid (9.02 mg/g).

Similarly, chemically modified CAC show an exceptional COD removal for real wastewater. Both CAC oxidized by nitric and sulfuric acids with optimum dose of 2.0 g/L and 3.0 g/L respectively, can significantly reduce the initial COD value of the real wastewater (250 mg/L) up to 100% and 90%, which was less than the permitted effluent limit of COD.

4.3. COLUMN STUDIES

Isotherm study is necessary to provide useful information on the practical utility and effectiveness of an adsorbent for metal removal. Therefore, fixed-bed column studies for all types of adsorbent using real wastewater were also conducted.

In column studies, regeneration of spent adsorbent and recovery of adsorbate material are the key factors in wastewater treatment application. To design such adsorption/desorption process in column operations, the adsorption capacities and adsorption kinetics between adsorbent and adsorbate need to be clearly defined. One way to obtain these characteristics is by examining the concentration of adsorbate in the effluent versus the number of bed volume (BV), which could be treated by the adsorbent until reaching a complete exhaustion. By using a breakthrough technique, the behavior of metal adsorption on the adsorbent surface can be evaluated.

4.3.1. Coconut Shell Charcoal (CSC)

4.3.1.1. Breakthrough

In this study, the similar conditions of column operations using synthetic wastewater were also imposed on real wastewater, except the Cr concentration of influent (8 mg/L), the initial pH of real wastewater (pH≈7.27), and the influent characteristics, which contains other metal contaminants and other organic impurities (Table 4.6). The breakthrough curves of all types of CSC for Cr removal for the first run are presented in Figure 4.32.

Figure 4.32 indicates that about 576 BV, corresponding to 25.2 L of influent, could be passed through the virgin column of the nitric-treated CSC without being detected in the effluent, while the breakthrough point of 1 mg/L of Cr concentration occurred at 864 BV (about 39.4 L of influent), which required a remarkably longer time than that of sulfuric-treated CSCCC and sulfuric-treated CSC (480 BV and 384 BV) respectively.

Exhaustion of the column, defined as the point where the breakthrough curve intersects the $C_e/C_o=1$, occurred at 1528 BV for the nitric-treated CSC. It is interesting to note that when the column was completely exhausted, the effluent pH (pH≈2.32) was found to be remarkably lower than the initial influent pH (pH≈7.27); thus, suggesting that pH played major roles on the removal capability of adsorbent during column operation. This indicates that a lower pH inside the column caused an increase in the columbic forces between the negative surface charge of CSC and Cr^{3+}, resulting in a higher Cr adsorption.

According to the first two regeneration cycles, it was found that the removal of the column on Cr tended to deteriorate with an increasing column cycle; thus, suggesting a lower adsorption of Cr in subsequent cycles. This could be attributed to the reduction in the number of available adsorption sites occupied by the residual Cr from the previous run.

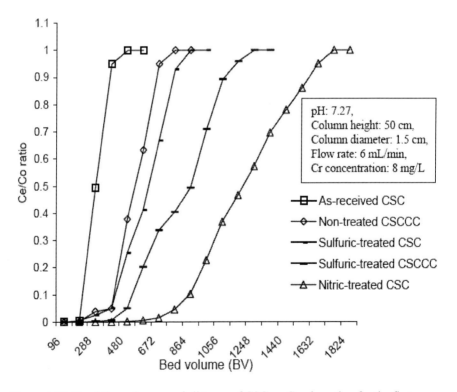

Figure 4.32. Breakthrough curve of all types of CSC on Cr adsorption for the first run.

Table 4.6. Comparison of the Cr adsorption capacity of all types of CSC between column and batch studies for the first run

No.	Types of CSC	Cr adsorption capacity (mg/g)		Difference of Cr adsorption capacity (%)
		Column studies	Batch studies	
1.	As-received CSC	3.31	3.22	2.72
2.	Non-treated CSCCC	6.26	5.02	19.81
3.	Sulfuric-treated CSC	6.95	5.18	25.46
4.	Sulfuric-treated CSCCC	9.55	8.32	12.88
5.	Nitric-treated CSC	13.90	13.69	1.51

Based on the breakthrough curve data, the Cr adsorption capacity of the nitric-treated CSC was found to be 13.90 mg/g in the first cycle, which was higher than that of the sulfuric-treated CSCCC and the sulfuric-treated CSC (9.55 mg/g and 6.95 mg/g) respectively. The Cr column capacities of all types of CSC on real wastewater, presented in Table 4.6, were compared to those of batch studies at the same Cr concentration of 8 mg/L.

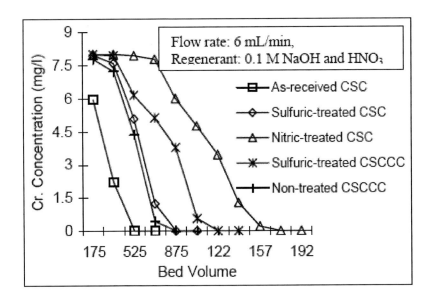

Figure 4.33. Regeneration curve of all types of CSC by 0.1 M NaOH solution.

Table 4.6 suggests that the Cr adsorption capacity of all types of CSC in column operation is higher than that of batch studies at the same Cr concentration of 8 mg/L. This could be attributed to the fact that in batch studies, the concentration gradient decreased with an increasing contact time; while in column operations, however, the adsorbent continuously had physico-chemical interactions with fresh feeding solutions at the interface of the adsorption zone, as the adsorbate solution passed through the column. Consequently, more Cr was adsorbed on the CSC surface in column operation than that in batch studies.

4.3.1.2. Regeneration

A good adsorbent, in addition to its high adsorption capacity on certain metal, should exhibit also a good regeneration for multiple uses. For this reason, the regenerability of CSC was tested by a series of desorption experiments using 0.1 M NaOH. The findings are presented in Figure 4.33.

Figure 4.33 suggests that a complete desorption of Cr occurred by NaOH solution. The recovery of Cr about 99.64% was obtained by desorption after the first regeneration of nitric-treated CSC (Table 4.7).

During the regeneration of spent nitric-treated CSC, column suffered a loss in Cr adsorption capacity (Table 4.7). It was observed that the column lost its capacity by approximately 0.65% after the first cycle, suggesting that Cr presence on the carbon surface of CSC from previous run did not adversely affect the removal performance of the column. Table 4.8 also suggests that the same column of nitric-treated CSC could be used for subsequent cycles.

Table 4.7. Comparison of total mass balance of Cr adsorbed on carbon before and after regeneration for the first cycle of all types of CSC

No.	Types of CSC	Cr before Regeneration (mg/g)	Cr after Regeneration (mg/g)	Regeneration Efficiency (%)*
1.	As-received CSC	3.31	3.20	96.68
2.	Non-treated CSCCC	6.26	6.22	99.36
3.	Sulfuric-treated CSC	6.95	6.88	98.99
4.	Sulfuric-treated CSCCC	9.55	9.50	99.48
5.	Nitric-treated CSC	13.90	13.85	99.64

*Remarks: % regeneration efficiency (RE) was calculated using Equation 3.3.

Table 4.8. Summary of column performance for
Cr adsorption by all types of CSC

Type of CSC	BV treated for the 1st run		Initial Cr adsorption capacity (mg/g)	Cr adsorption capacity (mg/g) at second run	Loss of adsorption capacity (%)*
	At breakthrough	At exhaustion			
As-received CSC	192	384	3.31	3.13	5.44
Non-treated CSCCC	288	672	6.26	6.22	0.64
Sulfuric-treated CSC	384	768	6.95	6.86	1.29
Sulfuric-treated CSCCC	480	1344	9.55	9.46	0.94
Nitric-treated CSC	864	1528	13.90	13.81	0.65

*Remarks: %loss of adsorption capacity was calculated using Equation 3.4.

Overall, the results of column experiment validated the preliminary findings in batch studies that nitric-treated CSC was an outstanding adsorbent for Cr removal at pH ranging from 6.5 to 7.5.

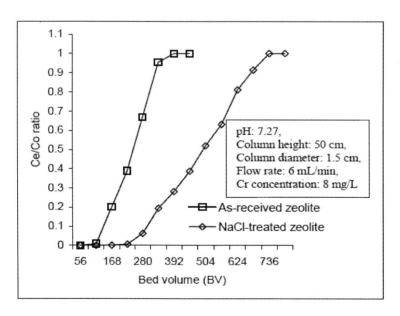

Figure 4.34. Breakthrough curves for as-received zeolite and NaCl-treated zeolite for the first run.

4.3.2. Zeolite

4.3.2.1. Breakthrough

To assess their effectiveness for Cr removal, column studies for as-received zeolite and NaCl-treated zeolite were also performed. Typical breakthrough curves, representing the ratio of effluent concentration (C_e) over influent concentration (C_o) versus the number of bed volume (BV) passed through the column until complete exhaustion, are presented in Figure 4.34.

Figure 4.34 reveals that almost complete removal of Cr from the influents at the initial stage of adsorption treatment was obtained before 224 BV (about 9.6 L of influent) had passed through the column. The breakthrough point of the NaCl-treated zeolite was found to occur at 280 BV, corresponding to 13.8 L for the first run. Compared to the former, as-received zeolite achieved earlier breakthrough point of 1 mg/L of Cr concentration at 147 BV and became completely exhausted at 392 BV. This remarkable difference in terms of Cr removal performance between the two types of zeolite suggests that chemical pretreatment of zeolite with NaCl significantly affected the course ion exchange in its surface to occur. Since the Na^+ of zeolite was in homoionic form before ion exchange, it could be replaced by Cr^{3+} in the solutions (at acidic conditions).

The adsorption capacities of zeolite, presented in Table 4.8, were calculated based on the breakthrough curve area under complete exhaustion. For comparison, the adsorption capacities, were determined from the adsorption equilibrium at an initial concentration of 8 mg/L, are also presented in the Table 4.9.

Although the Cr column capacities of as-received zeolite and NaCl-treated zeolite were slightly higher than those of batch studies, it is suggested that these results were comparable, thus indicating that for a field application, zeolite beds in series should be used. When running column in series, it is important to note that the first column should be run until attaining complete exhaustion ($C_e/C_o=1$).

Monitoring of the effluent pH during the adsorption cycle indicates a significant increase of pH from 7.27 to 9.02 during the first cycle, suggesting that Cr adsorption on the zeolite surface releases OH^- into the system.

4.3.2.2. Regeneration

In order to make treatment of Cr-contaminated water economically attractive, it is necessary to regenerate and reuse spent adsorbent after it has become completely saturated with adsorbate.

Therefore, Cr desorption from the zeolite surface was conducted with 0.1 *M* NaOH solution. The results are presented in Figure 4.35.

Table 4.9. Comparison of the Cr adsorption capacity of all types of zeolite between column and batch studies for the first run

No.	Types of zeolite	Cr adsorption capacity (mg/g)		Difference of Cr adsorption capacity (%)
		Column studies	Batch studies	
1.	As-received zeolite	3.82	3.06	19.89
2.	NaCl-treated zeolite	6.18	5.97	3.40

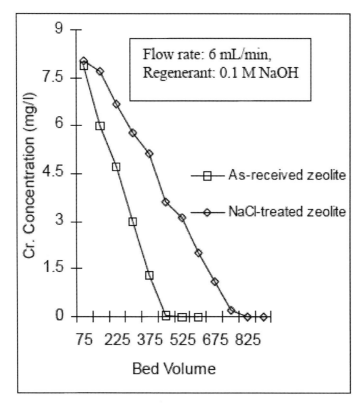

Figure 4.35. Regeneration curve of the two types of zeolite using 0.1 M NaOH.

It was found that the first 38.1 L of 0.1 M NaOH solution accounted for the recovery of 87% of chromium from the NaCl-treated zeolite (Table 4.10). This suggests that NaOH solution is suitable chemical to regenerate this adsorbent. Unlike the regeneration of CSC and/or CAC, HNO_3 solution was not passed through the zeolite column after the first Cr desorption with NaOH.

It is observed from Table 4.11 that the column performance of the NaCl-treated zeolite and/or as-received zeolite tended to deteriorate about 20-25% after the first two runs. The rate of this deterioration decreased with an increasing number of successive cycle could be attributed to the exposure to the highly alkaline regenerant solution (Ouki et al., 1993). In spite of this fact, it is suggested that NaCl-treated zeolite was also a reasonable low-cost adsorbent for Cr removal, as it could exhibit a reasonable Cr adsorption capacity (6.18 mg/g) for the first run and might have potential advantages over other chemical treatment methods such as activated sludge.

Table 4.10. Comparison of total mass balance of Cr adsorbed on zeolite before and after regeneration for the first cycle of all types of zeolite

No.	Types of zeolite	Cr before regeneration (mg/g)	Cr after regeneration (mg/g)	Regeneration efficiency (%)*
1.	As-received zeolite	3.82	3.01	78.79
2.	NaCl-treated zeolite	6.18	5.37	86.89

* Remarks: % regeneration efficiency (RE) was calculated using Equation 3.3.

Table 4.11. Summary of column performance for Cr adsorption by all types of zeolite

Type of zeolite	BV treated for the first run		Initial Cr adsorption capacity (mg/g)	Cr adsorption capacity (mg/g) at second run	Loss of adsorption capacity (%)*
	At break-through	At exhau-stion			
As-received zeolite	147	392	3.82	2.86	25.13
NaCl-treated zeolite	280	736	6.18	4.95	19.90

* Remarks: %loss of adsorption capacity was calculated using Equation 3.4.

4.3.3. Commercial Activated Carbon (CAC)

4.3.3.1. Breakthrough

Column operations for all types of CAC using real wastewater were also performed. The influent with an initial Cr concentration of 8 mg/L was continuously passed through a column packed with 27.0 g of CAC at a flow rate of 6.0 mL/min and an initial pH of 7.27. All breakthrough curves of all types of CAC for Cr removal are illustrated in Figure 4.36.

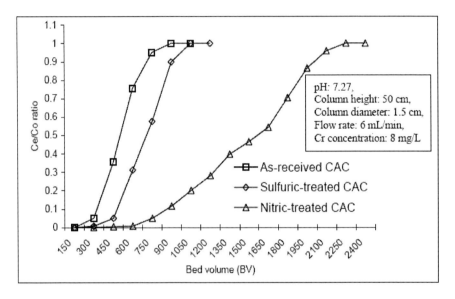

Figure 4.36. Breakthrough curve of all types of CAC on Cr adsorption.

Figure 4.36 presents that no Cr was detected in the effluent of the virgin nitric-treated CAC before 600 BV of influent (about 24.6 L) had passed through, while the breakthrough point of 1 mg/L in the effluent occurred at 750 BV, corresponding to 29.8 L. The amount of BV, which could be treated by the virgin nitric-treated CAC before achieving breakthrough point was remarkably larger compared to that of sulfuric-treated CAC and as-received CAC (450 and 300 BV), respectively. The Cr adsorption capacity of nitric-treated CAC, calculated from the breakthrough at complete exhaustion, was found to be 18.39 mg/g, which was significantly higher than other types of CAC (Table 4.12).

Table 4.12. Comparison of the Cr adsorption capacity of all types of CAC between column and batch studies for the first run

No.	Types of CAC	Cr adsorption capacity (mg/g)		Difference of Cr adsorption capacity (%)
		Column studies	Batch studies	
1.	As-received CAC	6.34	6.04	4.73
2.	Sulfuric-treated CAC	9.68	9.02	6.82
3.	Nitric-treated CAC	18.39	14.96	18.65

The monitoring of the effluent pH shows a significant decrease of pH from 7.27 to 2.51 during the first run of nitric-treated CAC. This indicates that the rapid decrease in column pH directly corresponds to the increase in C_e/C_o, as more adsorption of Cr occurred due to the decreasing column pH. Since adsorption was the dominant mechanism of Cr removal by nitric-treated CAC, a decrease in column capacity with increasing runs would be expected, considering that the number of its unoccupied adsorption sites decreased.

4.3.3.2. Regeneration

Regeneration of spent adsorbent is one of the most important aspects in wastewater treatment application, as its purpose is to return the adsorbent to its original condition for multiple uses and to restore its adsorption capacity without losing its chemical structure. It is important to note that the usefulness of a column is related to how long the bed will be last before renewal or regeneration become quite necessary.

After complete exhaustion, NaOH solution was used to desorb Cr from the column at the same flow rate of 6.0 mL/min. Chromium concentration in the effluent was determined and the regenerability of all types of CAC for the first cycle is presented in Figure 4.37.

Table 4.13 presents that the Cr recovery by the nitric-treated CAC was found to be 98.48%, which was slightly higher than the sulfuric-treated CAC (97.23%) and the as-received CAC (97.94%). It is interesting to note that Cr desorption by the nitric-treated CAC was found to proceed slowly and that an almost complete desorption of Cr could be eluted with 132.0 L of 0.1 M NaOH for the first regeneration, while about 41.4 L and 54 L of the same regenerant were required to desorp Cr from the as-received CAC and the sulfuric-treated CAC respectively. This suggests that more volume of the chemical was required to completely desorp Cr from the surface of nitric-treated CAC than the latter, resulting in a higher operational cost.

Table 4.14 reveals that the nitric-treated CAC lost its capacity by 0.98 % after the first cycle. At second exhaustion, its Cr adsorption capacity was found to be 18.21 mg/g, which means that about 99.02% of the Cr adsorption capacity of the nitric-treated CAC remained after one regeneration cycle. Thus, suggesting that a combined regenerant of NaOH and HNO₃ solution could restore the lost adsorption capacity of the nitric-treated CAC to almost its original value. It is quite interesting to observe that a slight decrease in sorption efficiency took place after the first run in the virgin column of nitric-treated CAC Therefore, the same column can be chemically regenerated for subsequent cycles without dismantling it.

Overall, this study clearly suggests that nitric-treated CAC was a promising adsorbent for Cr removal from contaminated water, as it has a high Cr adsorption capacity. Due to its outstanding performance in treating Cr-contaminated wastewater and the ease of column regeneration, further use of nitric-treated CAC column to treat other metal-bearing waste streams should be considered.

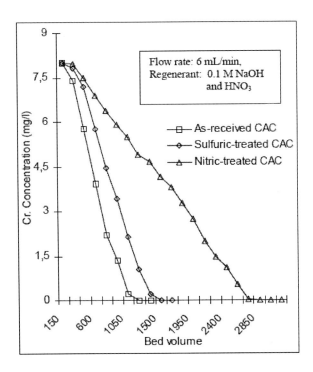

Figure 4.37. Regeneration curve of all types of CAC using 0.1 M NaOH.

Table 4.13. Comparison of total mass balance of Cr adsorbed on carbon before and after regeneration for the first cycle of all types of CAC

No.	Types of CAC	Cr before regeneration (mg/g)	Cr after regeneration (mg/g)	Regeneration efficiency (%)*
1.	As-received CAC	6.34	6.21	97.94
2.	Sulfuric-treated CAC	9.68	8.44	97.23
3.	Nitric-treated CAC	18.39	18.11	98.48

* Remarks: % regeneration efficiency (RE) was calculated using Equation 3.3.

Table 4.14. Summary of column performance for Cr adsorption by all types of CAC

Type of CAC	BV treated at the first run		Initial Cr adsorption capacity (mg/g)	Cr adsorption capacity (mg/g) at second run	Loss of adsorption capacity (%)*
	At breakthrough	At exhaustion			
As-received CAC	300	900	6.34	6.16	2.84
Sulfuric-treated CAC	450	1050	9.68	9.60	0.83
Nitric-treated CAC	750	2250	18.39	18.21	0.98

*Remarks: %loss of adsorption capacity was calculated using Equation 3.4.

4.3.4. Comparison of the Cr Adsorption Capacities of CSC, Zeolites, and CAC Using Real Wastewater in Column Studies

Similarly, the Cr adsorption capacities of all types of adsorbents in real wastewater are also evaluated and compared. Results are presented in Figure 4.38.

It was evident from Figure 4.38 and the column studies that the chemically modified CSC demonstrated remarkably better Cr adsorption capacity than as-received CSC (3.31 mg/g). Of those CSC in modified forms, the nitric-treated CSC has the highest Cr column capacity (13.90 mg/g) than the sulfuric-treated CSC and coated with chitosan (sulfuric-treated CSCCC) and the sulfuric-treated CSC with Cr adsorption capacities of 9.55 mg/g and 6.95 mg/g respectively. It is interesting to note that the nitric-treated CSC could achieve a breakthrough point of 1 mg/L of Cr concentration until 39.4 L of influent (864 BV) had passed through the column, thus suggesting that this adsorbent required a longer time to reach a breakthrough point than other types of CSC. This could be due to the fact that the oxidative pretreatment of CSC with nitric acid remarkably increased its negative surface charge to have electrostatic interactions with Cr^{3+} and generated more surface oxygen complex on the CSC surface (Aggarwal et al., 1999). As a result, the oxidized CSC had more available surface sites for Cr adsorption (at acidic conditions).

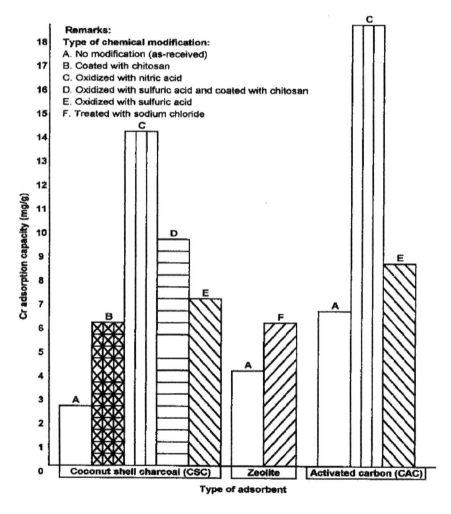

Figure 4.38. Comparison of the column capacity of CSC, zeolite, and CAC on chromium.

Similarly, zeolite treated with NaCl has a higher Cr adsorption capacity (6.18 mg/g) compared to as-received zeolite (3.82 mg/g). Pretreatment of zeolite with NaCl was found to enable the zeolite to treat more bed volume (280 BV) at a breakthrough point of 1 mg/L of Cr concentration for the first run than as-received zeolite (147 BV). This could be explained due to the fact that chemical pretreatment of zeolite with sodium chloride rendered the zeolite in the homoionic form of Na^+. Consequently, the Na^+ of zeolite could replace Cr^{3+} in the solution through ion exchange. Although the removal performance of NaCl-treated zeolite on Cr was not as high as that of the nitric-treated CSC, it was pointed out that this type of adsorbent has a reasonable Cr adsorption capacity (6.18 mg/g), which is fairly comparable to that of non-treated CSCCC (6.26 mg/g).

Compared to CSC and zeolite in as-received form (3.31 and 3.82 mg/g), as-received CAC has the most outstanding Cr adsorption capacity (6.34 mg/g). In fact, the nitric-treated CAC has nearly three times higher Cr column capacity (18.39 mg/g) than as-received CAC due to the fact that nitric acid is a strong oxidizing agent. Therefore, it could remarkably generate more surface sites on the surface of CAC for metal adsorption.

It is important to note also that no Cr was detected in the treated effluents of virgin nitric-treated CAC before 600 BV, which is corresponding to 24.6 L of influent. The breakthrough point of 1 mg/L of Cr concentration occurred at 750 BV (about 29.8 L of influent). This suggests that the nitric-treated CAC demonstrated the most outstanding Cr removal performances among other types of adsorbents in this study. For this reason, the utilization of the nitric-treated CAC for Cr removal from electroplating wastewater is recommended. However, considering its cost effectiveness and technical feasibility, nitric-treated CSC is preferable to the nitric-treated CAC for Cr removal, as its current commercial price is one third of that of CAC.

Overall, this present study demonstrates the removal capability of CSC, zeolite, and CAC for Cr removal from real Cr wastewater. Their local availability, sludge free operation, low chemical consumption, and reasonable Cr adsorption capacity are expected to reduce their operational cost during column operations and render this removal technique economically attractive for possible industrial applications.

CONCLUSION

Overall, it has been concluded from the present studies that surface modifications of CSC and CAC significantly have improved their Cr adsorption capacities. At the same Cr concentration (20 mg/L), among all types of surface modifications, CSC treated with nitric acid stands out for having the highest Cr adsorption capacity (10.88 mg/g) compared to other types of CSC such as sulfuric-treated CSC coated with chitosan (8.95 mg/g), CSC treated with sulfuric acid (4.05 mg/g), and CSC coated with chitosan (3.65 mg/g). Furthermore, it was also found that CSC in as-received form was highly efficient for Cr removal at lower concentrations (5 mg/L and 10 mg/L) with its adsorption capacities of 4.30 mg/g and 3.74 mg/g, respectively; this suggests that the adsorption capacity of CSC at 5 mg/L of Cr concentration was comparable to that of as-received CAC (4.72 mg/g) at 20 mg/L of Cr concentration.

Like CSC, CAC oxidized with nitric acid also demonstrated the most outstanding Cr adsorption capacity (15.47 mg/g) compared to CAC oxidized with sulfuric acid (8.94 mg/g) and as-received CAC (4.72 mg/g). In fact, in order to achieve a maximum Cr removal, nitric-oxidized CAC required a smaller amount of dose (2 g/L), a lower agitation speed (130 rpm), and less contact time (60 min) compared to as-received CAC (6 g/L of dose, 150 rpm of agitation speed, and 90 min of contact time respectively). These facts are very interesting, as the dose and contact time required for accomplishing a complete Cr removal are important parameters to achieve an economical wastewater treatment plant and to reduce cost inefficiency in the daily operation of the electroplating industry.

Although the CSC, zeolite, and CAC are suitable for treating Cr-rich effluent, it is important to note that their Cr adsorption capacities vary, depending on the physico-chemical characteristics of the individual adsorbent, the extent of surface modification, and the initial Cr concentration. Due to economic and technical considerations, such as the simplicity of the required adsorption system and the ability of an adsorbent to accept a wide variation of effluent Cr concentration, it is suggested that CSC is preferable to CAC and zeolite for Cr removal from electroplating wastewater.

Last, but not least, since the chemically modified CSC had been capable of performing an efficient and effective Cr removal, it could be expected that not only the local electroplating industries, but the living organisms and the surrounding environment benefited also from potential metal toxicity due to chromium. Therefore, the use of CSC for Cr removal could be technically feasible, and economically viable for water treatment application, thus contributing to the sustainability of the environment. All these advantages add more credits to coconut shell material for Cr(VI) removal from contaminated wastewater.

(Reprinted in part from *Chemosphere 54(7):951-967, Babel and Kurniawan, Cr(VI) removal from synthetic wastewater using coconut shell charcoal and commercial activated carbon modified with oxidizing agents and/or chitosan,* Copyright (2004), with permission from The Elsevier Publisher)

Chapter 5

CONCLUSION AND RECOMMENDATIONS

ABSTRACT

To bridge the existing research gaps in the field of wastewater treatment, this study has investigated and compared the treatment performance of several types of low cost adsorbents such as coconut shell charcoal (CSC) and zeolite as well as commercial activated carbon (CAC) in as-received and chemically modified forms for the removal of Cr(VI) from contaminated water. In accordance with the main objectives of the present study, this research has analyzed and assessed the treatability of real wastewater samples collected from a local electroplating industry in Rangsit (Thailand). In addition to various physico-chemical characteristics of the real wastewater samples, the ability of treated effluents to meet the increasingly strict limits imposed by local environmental legislation was investigated.

This chapter summarizes major findings of the present study in relation to the research objectives presented in Section 1.3 (Chapter 1) and makes a series of various recommendations for future work. It is divided into two sections. The first section presents encouraging results obtained from a series of batch and column studies using various types of adsorbents in as-received and modified forms for treatment of Cr(VI)-contaminated water. The second section presents various recommendations that will give readers insights how to follow up and apply the results of this project. It is also anticipated that this study will provide useful ideas for local environmental agencies to formulate and implement integrated strategies for water pollution control to enhance the sustainability of local environment, as envisioned by the UN Millennium Development Goals (MDG).

5.1. CONCLUSION

Based on the results of the studies presented in Chapter 4, a number of conclusions are drawn regarding the removal performance of all types of adsorbents employed for treatment of Cr(VI)-contaminated water in batch and column modes. The conclusions are summarized in order according to the objectives presented in Section 1.3 (Chapter 1).

5.1.1. Batch Studies

1. The first objective of this study was to find out the optimum conditions required to achieve a maximum removal of all adsorbents for chromium in batch studies. The conclusions of this objective are highlighted as follows:

Coconut Shell Charcoal
a. CSC in as-received and chemically modified forms was technically feasible and applicable for the removal of Cr(VI). Under optimum conditions (dose: 12 g/L; pH: 4.0; agitation speed: 150 rpm; contact time: 180 minutes), the HNO_3-treated CSC demonstrated a significantly higher removal performance (85%) than that treated with H_2SO_4 (65%) or coated with chitosan (60%) at the same initial concentration of 20 mg/L;
b. In spite of the outstanding removal of the HNO_3-treated CSC, the concentration of Cr(VI) in the treated effluents was still higher than 0.25 mg/L, the effluent limit imposed by local environmental legislation. This suggests that further treatments using biological processes such as activated sludge are required to comply with the permitted discharge standard;
c. The physico-chemical treatment applied to real Cr wastewater using nitric-oxidized CSC and/or nitric-treated CAC achieved a COD removal efficiency around 95% and 100% respectively with the optimum conditions as follows: dose= 7.5 g/L, pH= 4.5 for nitric-oxidized CSC and dose= 2.0 g/L, pH= 4.5 for nitric-treated CAC;

Zeolite
c. At the same initial Cr concentration of 20 mg/L, the NaCl-treated zeolite attained a higher Cr removal of 74% than did the as-received zeolite (57%) with the operating conditions as follows: the dose of treated zeolite: 9 g/L; pH: 4.5, agitation speed: 150 rpm, contact time: 180 minutes;

d. Although the treated zeolite could treat the Cr wastewater at an initial concentration of 20 mg/L, the treated effluent was still unable to meet the required Cr limit of less than 0.05 mg/L imposed by the US EPA;

Commercial Activated Carbon (CAC)

e. Under optimized conditions (2 g/L of dose, pH 4, 150 rpm of agitation speed and 180 min of contact time), adsorption treatment using nitric-modified CAC in batch studies was still ineffective to treat Cr(VI)-contaminated wastewater at an initial concentration of 20 mg/L; and

f. However, the removal of the nitric-modified CAC was still higher 98% than that of CAC treated with sulfuric acid (98%) and/or in as received form (90%) respectively at the same Cr concentrations of 20 mg/L.

2. The second objective was to assess and compare the adsorption capacity of all types of adsorbents using synthetic and real Cr wastewater in batch studies with varying Cr concentrations (from 5 to 10 mg/L). The conclusions of this objective are presented as follows:

Coconut Shell Charcoal

a. Compared to the Freundlich model, the Langmuir isotherm was slightly more representative for all types of CSC in removing target metal, as indicated by its higher coefficient correlations;

b. Nitric-treated CSC was found to be able to adsorb 10.88 mg of Cr, while as-received CSC adsorbed less the metal per gram (2.18 mg/g);

c. Both CSC and CAC, which have been chemically modified with nitric acid, have higher Cr adsorption capacities (CSC: 10.88 mg/g, CAC: 15.47 mg/g) than those of oxidized with sulfuric acid (CSC: 4.05 mg/g, CAC: 8.94 mg/g) and those of coated with chitosan (non-treated CSCCC: 3.65 mg/g) respectively;

d. The Cr adsorption capacities of all as-received adsorbents on real Cr wastewater at 8 mg/L of Cr concentration (CSC: 3.22 mg/g, CAC: 6.04 mg/g, zeolite: 3.06 mg/g) was comparable to those of synthetic wastewater at 10 mg/L of Cr concentration (CSC: 3.74 mg/g, CAC: 6.66 mg/g, zeolite: 3.55 mg/g);

Zeolite

d. Chemical pretreatment of zeolite with NaCl improved its Cr adsorption capacities (as-received zeolite: 1.79, NaCl-treated zeolite: 3.23 mg/g);

e. Like CSC, both types of zeolite were more applicable to simulate the Langmuir isotherm compared to the Freundlich model;

Commercial Activated Carbon (CAC)

f. As-received CAC had a significantly better Cr adsorption capacity (4.72 mg/g) than as-received CSC and/or as-received zeolite (CSC: 2.18 mg/g, zeolite: 1.79 mg/g) respectively at varying Cr concentrations from 5 to 10 mg/L;

g. The Freundlich isotherm was more representative than the Langmuir model for Cr adsorption by nitric-oxidized CAC (adsorption capacity: 15.47 mg/g); and

h. The removal efficiencies and adsorption capacities of all types of adsorbents on Cr vary, depending on the physical characteristics of the individual adsorbent, the extent of surface modification/treatment, and the initial concentration of Cr.

3. The third objective of the present study was to evaluate the ability of the treated effluents to meet the Cr limits imposed by environmental legislation (Thai PCD: 0.25 mg/L; US EPA: 0.05 mg/L). The conclusions of these objectives are summarized as follows:

Coconut Shell Charcoal (CSC) and Commercial Activated Carbon (CAC)

a. In spite of surface modifications with oxidizing agents such as sulfuric and/or nitric acids or coating with chitosan respectively, neither CSC nor CAC was able to generate treated effluents that comply with the strict requirements imposed by both the local environmental legislation or the US EPA;

Zeolite

b. Chemical pretreatment of zeolite with NaCl could not help the adsorbent meet the effluent discharge standards of Cr imposed by the Thai PCD and/or the US EPA.

4. The fourth objective was to study the regeneration for all types of spent CSC and CAC in column experiments using a combined regenerant of NaOH subsequently followed by HNO_3. The conclusion for this objective is summarized as follows:

a. Nitric-treated CSC and CAC were relatively excellent adsorbents for Cr removal, as they could exhibit multiple regenerations and high adsorption capacity after the first desorption cycles;

b. A combination of $0.1M$ NaOH and HNO_3 were found to be effective regenerants for both spent CSC and CAC as the same column was still effective after subsequent desorption cycles; and

c. Due to the inherent differences of nature, all types of adsorbent in column operation showed higher Cr adsorption capacities at the same Cr concentration of 20 mg/L than those in batch studies. In batch studies, the concentration gradient decreased with an increasing contact time; in column studies, however, the adsorbent continuously had physico-chemical interactions with fresh feeding solutions at the interface of the adsorption zone, as the adsorbate solution passed through the column.

5. The final objective of this study was to study the adsorption mechanisms of Cr(VI) by CSC and/or zeolite in the solutions. The conclusions for this objective are:

a. There are two consecutive adsorption mechanisms taking place during the metal removal: the first mechanism concerns with the reduction of Cr^{6+} to Cr^{3+}, and the second step controls the adsorption of Cr^{3+} on the surface of CSC. In this regard, the Cr adsorption is attributed to the electrostatic attractive interactions between the positive charge of Cr^{3+} and the negative surface charge of CSC; and

b. Unlike CSC, the mechanism of Cr removal by zeolite in the aqueous solution is facilitated by the ion exchange between Cr^{3+} and the Na^+ of zeolite.

5.2. RECOMMENDATIONS

Based on the results of the present studies, several recommendations are made to encompass new directions and strategies that can be adopted to improve water pollution control due to the generation of Cr(VI)-contaminated water caused by local electroplating industries. Therefore, keeping the results of the present study, various recommendations for future research directions are presented as follows:

5.2.1. Batch Studies

1. In-depth batch experimental studies should be conducted to evaluate the removal efficiency of adsorbent on total chromium (Cr^{6+} and Cr^{3+}). Theoretically, chromium compounds exist in the solutions in the form of hexavalent state. However, if these compounds, which may contain Cr^{3+}, are oxidized with $KMnO_4$ in the presence of H_2SO_4, the concentration of total chromium in the compounds can be determined. Then, the exact distribution of

removal efficiency between Cr^{6+} and Cr^{3+} will be known for each individual adsorbent;

2. Although 3 h of oxidation period for both CSC and CAC with nitric and/or sulfuric acids gave promising results of Cr removal efficiency, further investigations are required to increase the oxidation period up to 6 or 12 h, as it may enhance the removal efficiency and adsorption capacities of CSC and/or CAC on chromium;

3. Chemical modifications of CSC and/or CAC with different oxidizing agents such as ozone (O_3), ammonium persulfate ($[NH_4]_2S_2O_8$), and hydrogen peroxide (H_2O_2), need to be conducted to evaluate and find out stronger oxidizing agents, which may generate more surface oxygen complexes on their carbon surface;

4. Mechanisms of Cr adsorption and textural changes on the carbon surface of CSC and/or CAC due to oxidative pretreatment needs to be thoroughly investigated and evaluated;5. Since chromium, copper, nickel, and cadmium co-exist in real electroplating wastewater when they are discharged as plating wastes, a study needs to be performed to find out their interfering effects in metal-plating wastewater on Cr removal by the adsorbent;

6. As EDTA is a strong complexing agent and can form metal organic chelates with metal ions, a further investigation needs to be carried out to ascertain whether its presence in the electroplating wastewater may increase Cr removal by the adsorbent or not;

7. Because temperature may affect the nature of adsorption reaction, further investigations should be performed to find out whether Cr adsorption by the adsorbent is exothermic or endothermic. If the Langmuir constant (Q^o) decreases with an increase in temperature, this indicates that the adsorption is exothermic. However, if the 'Q^o' value improves with an increasing temperature, this suggests that the adsorption process is endothermic;

8. Since the Cr adsorption is attributed to the columbic forces between the negatively charged surface of the adsorbent and the positive charge of trivalent chromium ions at acidic conditions, it is recommended that further work should be conducted not only to find out the effects of ionic strength on adsorption, but also to evaluate whether an increase in ionic strength may enhance Cr removal by the adsorbent;

9. In this study, experiments were carried out using CSC and/or CAC in granular form ranging from 0.4 to 2.40 mm. It is suggested that further adsorption studies should be conducted with a smaller particle size to evaluate whether a reduction in the particle size of adsorbent increases or decreases its Cr removal by the adsorbent; and

10. Since the porosity of an adsorbent has significant effects on the removal of target metal, it is recommended that the porosity of CSC and CAC should be extended to enhance the effectiveness of Cr adsorption and the regeneration of spent adsorbents. The higher the porosity of an adsorbent has, the easier is the metal adsorption upon its surface and the longer is its service life.

5.2.2. Column Studies

1. In depth-column experimental studies should be conducted using a flow rate less than 6 mL/min. Since a lower flow rate of feeding solutions increases the physico-chemical interaction between adsorbent and adsorbate in column, the solution has more available residence time to diffuse into the carbon for adsorption before it is swept further through the column. Consequently, it may maximize the treated volume of feeding solutions until breakthrough, extend the life-span of the bed, and result in a higher Cr removal by the adsorbent;

2. To understand the adsorption behavior of chromium in terms of the distribution of charge species and the surface charge of adsorbent, it is recommended that a down-flow column operation should be performed at varying pH values;

3. In the present study, experiments were carried out using CSC and/or CAC in granular form ranging from 0.4 to 2.40 mm. It is suggested that the size of carbon particle needs to be significantly reduced to a finer size to increase its column performance. Using a finer particle range, adsorption breakthrough curve may be more efficient than that using a larger particle size, suggesting that a longer breakthrough time and a classic "S" shape profile of breakthrough curve may be achieved using a finer particle size;

4. To obtain more design data, it is suggested that bed depth-service time (BDST) analysis should be employed in column operations not only to calculate the effects of varying feed concentrations, effluents compositions, flow rate, but also to accurately predict the service time of an adsorbent in the column to the bed mass before regeneration or replacement of a spent adsorbent becomes necessary; and

5. Since the removal performance of CSC on Cr was satisfactorily efficient for industrial applications, a detailed cost-benefit analysis of using CSC needs to be conducted to evaluate the economic feasibility of its practical use in water treatment applications.

REFERENCES

Abia, AA; Horsfall, M; Didi, O. The use of chemically modified and unmodified cassava waste for the removal of Cd, Cu, and Zn ions from aqueous solution. *Biores. Technol.* 2003, 90(3), 345-348.

Abollino, O; Aceto, M; Malandrino, M; Sarzanini, C; Mentasti, E. Adsorption of heavy metals on Na-montmorillonite: Effect of pH and organic substances. *Water Res.*, 2003, 37, 1619-1627.

Aggarwal, D; Goyal, M; Basal, RC. Adsorption of chromium by activated carbon from aqueous solution. *Carbon*, 1999, 37, 1989-1997.

Ahmed, S; Chughtai, S; Keane, MA. The removal of cadmium and lead from aqueous solution by ion exchange with Na-Y zeolite. *Sep. Purif. Technol.*, 1998, 13, 57-64.

Ahn, KH; Song, KG; Cha, HY; Yeom, IT. Removal of ions in nickel electroplating rinse water using low-pressure nanofiltration. *Desalination*, 1999, 122, 77-84.

Ajmal, M; Rao, RAK; Ahmad, R; Ahmad, J. Adsorption studies on *Citrus reticulata* (fruit peel of orange) removal and recovery of Ni(II) from electroplating wastewater. *J. Hazard. Mater.*, 2000, 79, 117-131.

Ajmal, M; Rao, RAK; Ahmad, R; Ahmad, J; Rao, LAK. Removal and recovery of heavy metals from electroplating wastewater by using kyanite as an adsorbent. *J. Hazard. Maters.*, 2001, B87, 127-137.

Ajmal, M; Rao, RAK; Anwar, S; Ahmad, J; Ahmad, R. Adsorption studies on rice husk: removal and recovery of Cd(II) from wastewater. *Biores. Technol.*, 2003, 86, 147-149.

Akita, S; Castillo, LP; Nii, S; Takahashi, K; Takeuchi, H. Separation of Co(II)/Ni(II) via micelar-enhanced ultrafiltration using organophosporus acid extractant solubilized by nonionic surfactant. *J. Membr. Sci.*, 1999, 162, 111-117.

Aklil, A; Mouflih, M; Sebti, S. Removal of heavy metal ions from water by using calcined phosphate as a new adsorbent. *J. Hazard. Mater.*, 2004, 112(3), 183-190.

Alaerts, GJ; Jitjaturunt, V; Kelderman, P. Use of coconut shell-based activated carbon for chromium(VI) removal. *Water Sci. Technol.*, 1989, 21(12), 1701-1704.

Al-Degs, Y; Khraisheh, MAM; Allen, SJ; Ahmad, MN. Effect of carbon surface chemistry on the removal of reactive dyes from textile effluent. *Water Res.*, 2000, 34(3), 927-935.

Ali, AA; Bishtawi, RE. Removal of lead and nickel ions using zeolite tuff. *J. Chem. Technol. Biotechnol.*, 1997, 69, 27-34.

Aliane, A; Bounatiro, N; Cherif, AT; Akretche, DE. Removal of chromium from aqueous solution by complexation-ultrafiltration using a water-soluble macroligand. *Water Res.*, 2001, 35(9), 2320-2326.

Álvarez-Ayuso, E; García-Sánchez, A. Removal of heavy metals from wastewaters by vermiculites. *Environ. Technol.*, 2003, 24, 615-625.

Álvarez-Ayuso, E; García-Sánchez, A; Querol, X. Purification of metal electroplating wastewaters using zeolites. *Water Res.*, 2003, 37(20), 4855-4862.

Alvarez-Vazquez, H; Jefferson, B; Judd, SJ. Membrane bioreactors vs conventional biological treatment of landfill leachate: a brief review. *J. Chem. Technol. Biotechnol.*, 2004, 79, 1043-1049.

Alves, MM; Beca, CGG; Carvalho, RG; Castanheira, JM; Pereira, MCS; Vasconcelos, LAT. Chromium removal in tannery wastewaters "polishing" by *Pinus sylvestris* bark. *Water Res.*, 1993; 27(8), 1333-1338.

Andrus, M.E. A review of metal precipitation chemicals for metal-finishing applications. *Metal Finishing*, 2000, 98(11), 20-23.

Annadurai, A; Juang, RS; Lee, DJ. Adsorption of heavy metals from water using banana and orange peels. *Water Sci. Technol.*, 2002, 47(1), 185-190.

Aoki, T; Munemori, M. Recovery of chromium(VI) from wastewater with iron(III) hydroxide-I. *Water Res.*, 1982, 16, 793-796.

Archundia, C; Bonato, PS; Rivera, JFL; Maciol, LC, Collins, KE, Collins, CH. Reduction of Cr(VI) in acid solutions. *Sci. Total Environ.*, 1993, 130/131, 231-236.

Arias, M; Barral, MT; Mejuto, JC. Enhancement of copper and cadmium adsorption on kaolin by the presence of humic acids. *Chemosphere*, 2002, 48, 1081-1088.

Arulanantham, A; Balasubramanian, N; Ramakrishna, TV. Coconut shell carbon for treatment of cadmium and lead containing wastewater. *Metal Finishing*, 1989, 87, 51-55.

Aung, N.N. *Adsorption/Desorption of Heavy Metals Using Chitosan*. Bangkok: Asian Institute of Technology; 1997.

Ayoub, GM; Semerjian, L; Acra, A; El Fadel, M; Koopman, B. Heavy metal removal by coagulation with seawater liquid bittern. *J. Environ. Eng.*, 2001, 127(3), 196-202.

Babel, S; Kurniawan, TA. Low-cost adsorbents for heavy metals uptake from contaminated water: a review. *J. Hazard. Mater.*, 2003a, 97(1-3), 219-43.

Babel, S; Kurniawan, TA. A research study on Cr(VI) removal from contaminated wastewater using natural zeolite. *J. Ion Exchange*, 2003b, 14, 289-292.

Babel, S; Kurniawan, TA. Chromium removal from electroplating wastewater using chemically treated zeolite. In: *Proc. the 9th World Filtration Congress*, New Orleans (USA); 18-22 April 2004; 2004a; pp. 1-14.

Babel, S; Kurniawan, TA. Cr(VI) removal from synthetic wastewater using coconut shell charcoal and commercial activated carbon modified with oxidizing agents and/or chitosan. Chemosphere, 2004b, 54(7), 951-967.

Babić, BM; Milonjic, SK; Polovina, MJ; Cupic, S, Kaludjerovic, BV. Adsorption of zinc, cadmium, and mercury ions from aqueous solutions on an activated carbon cloth. *Carbon*, 2002, 40, 1109-1115.

Bailey, SE; Olin, TJ; Bricka, M; Adrian, DD. A review of potentially low-cost sorbents for heavy metals. *Water Res.*, 1999, 33(11), 2469-2479.

Balkaya, N. Variation of pH, conductivity, and potential values in chromium(VI) removal by wool. *Environ. Technol.*, 2002, 23, 11-16.

Bansode, RR; Losso, JN; Marshall, WE; Rao, RM; Portier, RJ. Adsorption of metal ions by pecan shell-based granular activated carbons. *Biores. Technol.*, 2003, 89, 115-119.

Bayat, B. Comparative study of adsorption properties of Turkish fly ashes II. The case of chromium(VI) and cadmium(II). *J. Hazard. Maters.*, 2002, B95, 275-290.

Bayat, B. Comparative study of adsorption properties of Turkish fly ashes I. The case of nickel(II), copper(II), and zinc(II). *J. Hazard. Mater.*, 2002, B95, 251-273.

Benefield, LD; Morgan, JM. Chemical precipitation, In: *Water Quality and Treatment*; Letterman, RD.; Ed.; McGraw-Hill Inc: NY, 1999; pp 10.1-10.57.

Benito, Y; Ruiz, ML. Reverse osmosis applied to metal finishing wastewater. *Desalination*, 2002, 142, 229-234.

Bishnoi, NR; Bajaj, M; Sharma, N; Gupta, A. Adsorption of Cr(VI) on activated rice husk carbon and activated alumina. *Biores. Technol.*, 2003, 91(3), 305-307.

Blanchard, G; Maunaye, M; Martin, G. Removal of heavy metals from water by means of natural zeolites. *Water Res.*, 1984, 18(12), 1501-1507.

Blöcher, C; Dorda, J; Mavrov, V; Chmiel, H; Lazaridis, NK; Matis, KA. Hybrid flotation− membrane filtration processes for the removal of heavy metal ions from wastewater. *Water Res.*, 2003, 37, 4018-4026.

Bohdziewicz, J; Bodzek, M; Wąsik, E. The application of reverse osmosis and nanofiltration to the removal of nitrates from groundwater. *Desalination*, 1999, 121, 139-147.

Boonamuayvitaya, V; Chaiya, C; Tanthapanicchakoon, W; Jarudilokul, S. Removal of heavy metals by adsorbent prepared from pyrolyzed coffee residues and clay. *Sep. Purif. Technol.*, 2004, 35(1), 11-22.

Bose, P; Bose, MA; Kumar, S. Critical evaluation of treatment strategies involving adsorption and chelation for wastewater containing copper, zinc, and cyanide. *Adv. Environ. Res.*, 2002, 7, 179-195.

Brandhuber, P; Amy, G. Arsenic removal by a charged ultrafiltration membrane-influences of membrane operating conditions and water quality on arsenic rejection. *Desalination*, 2001, 140, 1-14.

Brown, P; Jefcoat, IA; Parrish, D; Gill, S, Graham, S. Evaluation of the adsorptive capacity of peanut hull pellets for heavy metals in solution. *Adv. Environ. Res.*, 2000, 4, 19-29.

Bruggen, B; Vandecasteele, C. Distillation vs membrane filtration: overview of process evolutions in seawater desalination. *Desalination*, 2002, 143, 207-218.

Bruggen, B; Vandecasteele, C. Removal of pollutants from surface water and groundwater by nanofiltration: overview of possible applications in the drinking water industry. *Environ. Pollut.*, 2003, 122, 435-445.

Business Economics Department. Whole Kingdom's Coconut Production. Thai Ministry of Commerce. 1998.

Calace, N; Nardi, E; Petronio, BM; Pietrolettti, M; Tosti, G. Metal ion removal from water by sorption on paper mill sludge. *Chemosphere*, 2003, 51(8), 797-803.

Candela, MP; Martinez, JM; Macia, RT. Chromium(VI) removal with activated carbon. *Water Res.*, 1995, 29(9), 2174-2180.

Chakir, A; Bessiere, J; Kacemi, KE; Marouf, M. A comparative study of the removal of trivalent chromium from aqueous solutions by bentonite and expanded perlite. *J. Hazard. Mater.*, 2002, B95, 29-46.

Chantawong, V; Harvey, NW; Bashkin, VN. Comparison of heavy metal adsorptions by Thai kaolin and ballclay. *Water, Air, Soil Pollut.*, 2003, 148, 111-125.

Charerntanyarak, L. Heavy metals removal by chemical coagulation and precipitation. *Water Sci. Technol.*, 1999, 39(10-11), 135-138.

Chen, GH. Electrochemicals technologies in wastewater treatment. *Sep. Purif. Technol.*, 2004, 38(1), 11-41.

Chen, JP; Chang, SY; Hung, YT. Electrolysis. In: *Physico-chemical Treatment Processes*; Wang, LK; Hung, YT; Shammas, NK.; Eds.; Humana Press: New Jersey, 2004; Vol.3, pp 359-378.

Chen, JP; Wang, XY. Removing copper, zinc, and lead ion by granular activated carbon in pretreated fixed-bed columns. *Sep. Purif. Technol.*, 2000, 19, 157-167.

Cheng, RC; Liang, S; Wang, H.C. Beuhler, MD. Enhanced coagulation for arsenic removal. *J. AWWA*, 1994, 86(9), 79-90.

Choi, H; Zhang, K; Dionysiou, DD; Oerther, DB; Sorial, GA. Effect of permeate flux and tangential flow on membrane fouling for wastewater treatment. *Sep. Purif. Technol.*, 2005, 45, 68-78.

Cotton, FA; Wilkilson, G. *Advanced Inorganic Chemistry*. New York: John Wiley and Sons; 1988.

CSIRO, 2004. Activated carbon. Available from:
<http://www.enecon.com.au/activ_carbon>

Curković, L; Stefanovic, SC; Filipan, T. Metal ion exchange by natural and modified zeolites. *Water Res.*, 1997, 31(6), 1379-1382.

Dąbrowski, A; Hubicki, Z; Podkościelny, P; Robens, E. Selective removal of the heavy metals from waters and industrial wastewaters by ion-exchange method. *Chemosphere*, 2004, 56(2), 91-106.

Dakiky, M; Khamis, M; Manassra, A; Mer'eb, M. Selective adsorption of Cr(VI) in industrial wastewater using low-cost abundantly available adsorbents. *Adv. Environ. Res.*, 2002, 6, 533-540.

Daneshvar, N; Salari, D; Aber, S. Chromium adsorption and Cr(VI) reduction to trivalent chromium in aqueous solutions by soya cake. *J. Hazard. Maters.*, 2002, B94, 49-61.

Dantas, TND; Neto, AAD; Moura, MCP. Removal of chromium from aqueous solutions by diatomite treated with microemulsion. *Water Res.*, 2001, 35(9), 2219-2224.

Dean, SA; Tobin, JM. Uptake of chromium cations and anions by milled peat. *Resource, Conservation & Recycling*, 1999, 27, 151-156.

Demirbaş, E; Kobya, M; Oncel, S; Sencan, S. Removal of Ni(II) from aqueous solution by adsorption onto hazelnut shell activated carbon: equilibrium studies. *Biores. Technol.* 2002, 84, 291-293.

Dimitrova, SV. Metal sorption on blast furnace slag. *Water Res.*, 1996, 30(1), 228-232.

Dobrevsky, I; Todorova-Dimova, M; Panayotova, T. Electroplating rinse wastewater treatment by ion exchange. *Desalination*, 1996, 108, 277-280.

Doyle, FM; Liu, ZD. The effect of triethylenetetraamine (trien) on the ion flotation of Cu^{2+} and Ni^{2+}. *J. Coll. Int. Sci.*, 2003, 258, 396-403.

Dries, J; Bastiaens, L; Springel, D; Agathos, SN; Diels, L. Combined removal of chlorinated ethenes and heavy metals by zerovalent iron in batch and continuous flow columns systems. *Environ. Sci. Technol.*, 2005, 39, 8460-8465.

Edwards, M. Chemistry of arsenic removal during coagulation and Fe-Mn oxidation. *J. AWWA,* 1994, 86(9), 64-78.

El-Hendawy, AA. Influence of HNO_3 oxidation on the structure and adsorptive properties of corncob-based activated carbon. *Carbon,* 2003, 41, 713-722.

Elimelech, M; O'Melia, CR. Kinetics of deposition of colloidal particles in porous media. *Environ. Sci. Technol.*, 1990, 24, 1528-1536.

Environmental Protection Agency (EPA), USA. Chemical Precipitation. US EPA, Washington DC, 2000. (EPA832-F-00-018)

Environmental Protection Agency (EPA), USA. *Control and treatment technology for the metal finishing industry, sulfide precipitation. Summary Report.* US EPA, Washington DC, 1980. (EPA-625/8-80/003)

Environmental Protection Agency (EPA), USA. *Development document for effluent limitations guidelines and standards for the metal finishing point source category.* US EPA, Washington DC, 1980. (EPA-440/1-80/091-a)

Environmental Protection Agency (US EPA), 2004. *National pollutant discharge elimination system (NPDES).* Available from: <http://www.cfpub.epa.gov/npdes/home.cfm?program_id=3>

Environmental Protection Department (EPD), Hong Kong. *Technical memorandum standards for effluents discharged into drainage and sewerage systems, inland and coastal water.* 2005. Available from: <http://www.legislationblis_ind.nsf/ e1bf50c09a33d3dc482564840019d2f4/034c887c9e2774078825648a005d 23b4?OpenDocument>

Environmental Protection Department (HK EPD), 2004. *Technical memorandum standards for effluents discharged into drainage and sewerage systems, inland and coastal water*. 2004. Available from: <http://www.legislationblis_ind.nsf/e1bf50c09a33d3dc482564840019d2f4/034c887c9e2774078825648a005d23b4?OpenDocument>

Faur-Brasquet, C; Kadirvelu, K; Le Cloirec, P. Removal of metal ions from aqueous solution by adsorption onto activated carbon cloths: adsorption competition with organic matter. *Carbon*, 2002, 40, 2387-2392.

Fei, C; Sheng, LG; Wei, YW; Jun, WY. Preparation and adsorption ability of polysulfone microcapsules containing modified chitosan gel. *Tsinghua Sci. Technol.*, 2005, 10(5), 535-541.

Feng, D; Van Deventer, JSJ; Aldrich, C. Removal of pollutants from acid mine wastewater using metallurgical by-product slags. *Sep. Purif. Technol.*, 2004, 40(1), 61-67.

Findon, A; McKay, G; Blair, HS. Transport studies for the sorption of copper ions by chitosan. *J. Environ. Sci. Health Part A*, 1993, 28, 173-185.

Gang, D; Banerji, SK; Clevenger, TE. Chromium(VI) removal by PVP-coated silica gel. Practice Periodical of Hazardouz, Toxic & Radioactive. *Waste Manage.*, 2000, 4(3), 105-110.

Ghosh, UC; Dasgupta, M; Debnath, S; Bhat, SC. Studies on management of chromium(VI)-contaminated industrial waste effluent using hydrous titanium oxide (HTO). *Water, Air, Soil Pollut.*, 2003, 143, 245-256.

Gode, F; Pehlivan. E. A comparative study of two chelating ion-exchange resins for the removal of chromium(III) from aqueous solution. *J. Hazard. Mater.*, 2003, 100(1-3), 231-243.

Goel, J; Kadirvelu, K; Rajagopal, C; Garg, VK. Removal of lead(II) by adsorption using treated granular activated carbon: Batch and column studies. *J. Hazard. Mater.*, 2005, B125, 211-220.

Goyal, M; Rattan, VK; Aggarwal, D; Bansal, RC. Removal of copper from aqueous solutions by adsorption on activated carbons. *Coll. Surf. A: Physicochem. & Eng. Aspects*, 2001, 190, 229-238.

Grebenyuk, VD; Sobolevskaya, TT; Machino, AG. Prospective of development of electroplating production effluent purification method. *Chem. Technol. Water*, 1989, 11 (5), 407-411.

Grebenyuk, VD; Sorokin, GV; Verbich, SV; Zhiginas, LH; Linkov, VM; Linkov, NA; Smit, JJ. Combined sorption technology of heavy metal regeneration from electroplating rinse waters. *Water SA*, 1996, 22(4), 381-384.

Gupta, DC; Tiwari, UC. Alumunium oxide as adsorbent for removal of hexavalent chromium from aqueous waste. *Ind. J. Environ. Hlth.*, 1985, 27(3), 205-215.

Gupta, VK. Equilibrium uptake, sorption dynamics, process development, and column operations for the removal of copper and nickel from aqueous solution and wastewater using activated slag, a low cost adsorbent. *Ind. Eng. Chem. Res.*, 1998, 37, 192-202.

Gupta, VK; Morhan, D; Sharma, S; Park, KT. Removals of chromium(VI) from electroplating industry wastewater using bagasse fly ash-a sugar industry waste material. *The Environmentalist*, 1999, 19, 129-136.

Gupta, VK; Jain, CK; Ali, I; Sharma, M; Saini, SK. Removal of cadmium and nickel from wastewater using bagasse fly ash-a sugar industry waste. *Water Res.*, 2003, 37, 4038-4044.

Hamadi, NK; Chen, XD; Farid, MM; Lu, MGQ. Adsorption kinetics for the removal of chromium(VI) from aqueous solution by adsorbents derived from used tyres and sawdust. *Chem. Eng. J.*, 2001, 84, 95-105.

Han, I; Schlautman, MA; Batchelor, B. Removal of hexavalent chromium from groundwater by granular activated carbon. *Water Environ. Res.*, 2000, 72, 29-39.

Hasan, S; Hashim, MA; Gupta, BS. Adsorption of NiSO$_4$ on Malaysian rubber-wood ash. *Biores. Technol.*, 2000, 72, 153-158.

Hasar, H. Adsorption of nickel(II) from aqueous solution onto activated carbon prepared from almond husk. *J. Hazard. Maters.*, 2003, B97, 49-57.

Hu, ZH; Lei, L; Lei, YJ; Ni, YM. Chromium adsorption on high performance activated carbons from aqueous solution. *Sep. Purif. Technol.*, 1998, 31, 13-18.

Huang, CP. The removal of chromium(VI) from dilute aqueous solution by activated carbon. *Water Res.*, 1977, 11, 673-679.

Huang, CP; Bowers, AR. The use of activated carbon for chromium(VI) removal. *Prog. Wat. Technol.*, 1978, 10(5), 45-64.

Huang, CP; Wu, MH. Chromium removal by carbon adsorption. *J. Water Pollut. Control Federation*, 1977, 47, 2437-2446.

Hurd, SM. Low pressure reverse osmosis membrane for treatment of landfill leachate, University of Carleton, Ottawa, 1999. (Master Thesis)

Ik, J; Zoltek, J. Chromium removal with activated carbon. *Prog. Wat. Technol.*, 1977, 9, 143-155.

Inbaraj, B; Sulochana, N. Carbonised jackfruit peel as an adsorbent for the removal of Cd(II) from aqueous solution. *Biores. Technol.*, 2004, 94(1), 49-52.

Inglezakis, VJ; Diamandis, NA, Loizidou, MD; Grigoropoulou, HP. Effects of pore clogging on kinetics of lead uptake by clinoptilolite. *J. Coll. Int. Sci.*, 1999, 215, 54-57.

Inglezakis, VJ; Papadeas, CD; Loizidou, MD; Grigoropoulou, H.P. Effects of pretreatment on physical and ion exchange properties of natural clinoptilollite. *Environ. Technol.*, 2001, 22, 75-82.

Itoi, S; Nakamura, I; Kawahara, T. Electrodialytic recovery process of metal finishing wastewater. *Desalination*, 1980, 32, 383-389.

Janssen, LJJ; Koene, L. The role of electrochemistry and electrochemical technology in environmental protection. *Chem. Eng. J.*, 2002, 85, 137-146.

Jokela, P; Keskitalo, P. Plywood mill water system closure by dissolved air flotation treatment. *Water Sci. Technol.*, 1999, 40(11-12), 33–42.

Juang, RS; Lin, SH; Wang, TY. Removal of metal ions from the complexed solutions in fixed bed using a strong acid ion exchange resin. *Chemosphere*, 2003, 53(10), 1221-1228.

Juang, RS; Shiau, RC. Metal removal from aqueous solutions using chitosan-enhanced membrane filtration. *J. Membr. Sci.*, 2000, 165, 159-167.

Jüttner, K; Galla, U; Schmieder, H. Electrochemical approaches to environmental problems in the process industry. *Electrochim. Acta*, 2000, 45, 2575-2594.

Kabay, N; Arda, M; Saha, B; Streat, M. Removal of Cr(VI) by solvent impregnated resins (SIR) containing aliquat 336. *React. Funct. Polym.*, 2003, 54, 103-115.

Kadirvelu, K; Namasivayam, C. Activated carbon from coconut coirpith as metal adsorbent: adsorption of Cd(II) from aqueous solution. *Adv. Environ. Res.*, 2003, 7, 471-478.

Kadirvelu, K; Senthilkumar, P; Thamaraiselvi, K; Subburam, V. Activated carbon prepared from biomass as adsorbent: elimination of Ni(II) from aqueous solution. *Biores. Technol.*, 2002, 81, 87-90.

Kadirvelu, K; Thamaraiselvi, K; Namasivayam, C. Adsorption of Ni(II) from aqueous solution onto activated carbon prepared from coirpith. *Sep. Purif. Technol.*, 2001b, 24, 497-505.

Kadirvelu, K; Thamaraiselvi, K; Namasivayam, C. Removal of heavy metals from industrial wastewaters by adsorption onto activated carbon prepared from an agricultural solid waste. *Biores. Technol.*, 2001a, 76, 63-65.

Karabulut, S; Karabakan, A; Denizli, A; Yürüm, Y. Batch removal of copper(II) and zinc(II) from aqueous solutions with low rank Turkish coals. *Sep. Purif. Technol.*, 2000, 18, 177-184.

Kaya, A; Oren, AH. Adsorption of zinc from aqueous solutions to bentonite. *J. Hazard. Mater.*, 2005, B125, 183-189.

Keane, MA. The removal of copper and nickel from aqueous solution using Y zeolite ion exchangers. *Coll. Surf. A*, 1998, 138, 11-20.

Ko, DCK; Porter, JF; McKay, G. Determination of solid-phase loading for the removal of metal ions from effluents using fixed-bed adsorbers. *Environ. Sci. Technol.*, 2001, 35, 2797-2803.

Kobya, M. Removal of Cr(VI) from aqueous solutions by adsorption onto hazelnut shell activated carbon: kinetic and equilibrium studies. *Biores. Technol.*, 2004, 91, 317-321.

Kocaoba, S; Akcin, G. Removal and recovery of chromium and chromium speciation with MINTEQA. *Talanta*, 2002, 57, 23-30.

Kongsricharoern, N. *Application of Electrochemical Precipitation of Treatment of Cr Wastewater.* Bangkok: Asian Institute of Technology; 1994.

Kongsricharoern, N; Polprasert, C. Chromium removal by a bipolar electrochemical precipitation process. *Water Sci. Technol.*, 1996, 34(9), 109-116.

Kongsricharoern, N; Polprasert, C. Electrochemical precipitation of chromium (Cr^{6+}) from an electroplating wastewater. *Water Sci. Technol.*, 1995, 31(9), 109-117.

Korngold, E; Belayev, N; Aronov, L. Removal of chromates from drinking water by anion exchangers. *Sep. Purif. Technol.*, 2003, 33(2), 179-187.

Kotaś, J.; Stasicka, Z. Chromium occurrence in the environment and methods of its speciation. *Environ. Pollut.*, 2000, 107, 263-283.

Krishnan, KA; Anirudhan, TS. Removal of cadmium(II) from aqueous solutions by steam-activated sulphurised carbon prepared from sugar-cane bagasse pith: kinetics and equilibrium studies. *Water SA.* 2003, 29(2), 147-156.

Kryvoruchko, A; Yurlova, L; Kornilovich, B. Purification of water containing heavy metal by chelating-enhanced ultrafiltration. *Desalination*, 2002, 144, 243-248.

Kurniawan, TA; Babel, S. A research study on Cr(VI) removal from contaminated wastewater using low-cost adsorbents and commercial activated carbon. In: *Proceedings of the 2nd International Conference on Energy Technology Towards a Clean Environment (RCETE)*, Phuket (Thailand); 12-14 February 2003; Vol. 2, pp.1110-1117.

Kurniawan, TA; Babel, S. Chromium removal from electroplating wastewater using low-cost adsorbents and commercial activated carbon. In:

Proceedings of the 5ᵗʰ International Summer Symposium, Tokyo (Japan); 26 July 2003; pp. 345-348.

Kurniawan, TA; Chan, G.; LO, W; Babel, S. Physico-chemical treatment techniques for treatment of wastewater laden with heavy metals. Chemical Engineering Journal, 2006a, 118(1-2): 83-98.

Kurniawan, TA; Chan, GYS; Lo, WH; Babel, S. Comparisons of low-cost adsorbents for treating wastewaters laden with heavy metals. *Sci. Total Environ.*, 2006b, 366(2-3): 407-424.

Laine, JM; Vial, D; Moulart, P. Status after 10 years of operation – overview of UF technology today. *Desalination,* 2000, 131, 17-25.

Lainé, S; Poujol, T; Dufay, S; Baron, J; Robert, P. Treatment of stormwater to bathing water quality by dissolved air flotation, filtration and ultraviolet disinfection. Water Sci. Technol., 1998, 38(10), 99–105.

Lakatos, J; Brown, SD; Snape, CE. Coals as sorbents for the removal and reduction of hexavalent chromium from aqueous wastestreams. *Fuel*, 2002, 81, 691-698.

Lalvani, SB; Wiltowski, T; Hubner, A; Weston A; Mandich, N. Removal of hexavalent chromium and metal cations by a selective and novel carbon adsorbent. *Carbon,* 1998, 36(7-8), 1219-1226.

Lazaridis, NK; Matis, KA; Webb, M. Flotation of metal-loaded clay anion exchangers. Part I: the case of chromate. *Chemosphere*, 2001, 42, 373-378.

Lee, CK; Low, SK; Kek, KL. Removal of chromium from aqueous solution. *Biores. Technol.*, 1995, 54, 183-189.

Lee, TY; Lim, HJ; Lee, YH; Park, JW. Use of waste iron metal for removal of Cr(VI) from water. *Chemosphere*, 2003, 53(5), 479-485.

Lee, TY; Park, JW; Lee, JH. Waste green sands as a reactive media for the removal of zinc from water. *Chemosphere*, 2004, 56, 571-581.

Leinonen, H; Lehto, J. Purification of metal finishing wastewaters with zeolites and activated carbons. *Water Manage. Res.*, 2001, 19, 45-57.

Leyva-Ramos, R. Rangel-Mendez, JR; Mendoza-Barron, J; Fuentes-Rubio, L; Guerrero-Coronado, RM. Adsorption of cadmium(II) from aqueous solution onto activated carbon. *Water Sci. Technol.*, 1997, 35(7), 205-211.

Leyva-Ramos, R; Jacome, LAB; Barron, JM; Rubio, LF; Coronado, RMG. Adsorption of zinc(II) from an aqueous solution onto activated carbon. *J. Hazard. Mater.*, 2002, B90, 27-38.

Li, YJ; Zeng, XP; Liu, YF, Yan, SS; Hu, ZH; Ni, YM. Study on the treatment of copper-electroplating wastewater by chemical trapping and flocculation. *Sep. Purif. Technol.*, 2003, 31, 91-95.

Lin, SH; Juang, RS. Heavy metal removal from water by sorption using surfactant-modified montmorrilonite. *J. Hazard. Mater.*, 2002, B92, 315-326.

Lin, SH; Kiang, CD. Chromic acid recovery from waste acid solution by an ion exchange process: equilibrium and column ion exchange modeling. *Chem. Eng. J.*, 2003, 92, 193-199.

Lisovskii, A; Semiat, R; Aharoni, C. Adsorption of sulfur dioxide by activated carbon treated by nitric acid: I. effect of preheating on the adsorption properties. *Carbon*, 1997; 35(10-11), 1645-1648.

Liu, MH; Zhang, HZ; Xin, SD; Yun, L; Wei-Guo; Zhan, HY. Removal and recovery of chromium(III) from aqueous solutions by a spheroidal cellulose adsorbent. *Water Environ. Res.*, 2001, 73(3), 322-328.

Lo, SL; Tao, YC. Economical analysis of waste minimization for electroplating plants. *Water Sci. Technol.*, 1997, 36(2-3), 383-390.

Low, KS; Lee, CK; Leo, AC. Removal of metals from electroplating wastes using banana pith. *Biores. Technol.*, 1995, 51, 227-231.

MacKay, G; Blair, HS; Findon, A. *Sorption of metal ions by chitosan*, in: Eccles H, Hunt S (Eds.), Immobilisation of ions by biosorption. Chichester (UK); Ellis Horwood; 1986.

Madaeni, SS; Mansourpanah, Y. COD removal from concentrated wastewater using membranes. *Filtr. Sep.*, 2003, 40, 40-46.

Marder, L; Bernardes, AM; Ferreira, JZ. Cadmium electroplating wastewater treatment using a laboratory-scale electrodialysis system. *Sep. Purif. Technol.*, 2003, 37(3), 247-255.

Marshall, WE; Wartelle, LH; Boler, DE; Johns, MM; Toles, CA. Enhanced metal adsorption by soybean hulls modified with nitric acid. *Biores. Technol.*, 1999, 69, 263-268.

Martin, RJ; Ng, WJ. The repeated exhaustion and chemical regeneration of activated carbon. *Water Res.*, 1987, 21(8), 961-965.

Martinez, SA; Rodriguez, MG; Aguolar, R; Soto, G. Removal of chromium hexavalent from rinsing chromating waters electrochemicals reduction in a laboratory pilot plant. *Water Sci. Technol.*, 2004, 49(1), 115-122.

Matis, KA; Zouboulis, AI; Gallios, GP; Erwe, T; Blöcher, C. Application of flotation for the separation of metal-loaded zeolite. Chemosphere, 2004, 55, 65-72.

Matis, KA; Zouboulis, AI; Lazaridis, NK; Hancock, IC. Sorptive flotation for metal ions recovery. *Int. J. Miner. Process*, 2003, 70, 99-108.

Mavros, P; Zouboulis, AI; Lazaridis, NK. Removal of metal ions from wastewaters. The case of nickel. *Environ. Technol.*, 1993, 14, 83-91.

Mavrov, V; Erwe, T; Blöcher, C. Chmiel, H. Study of new integrated processes combining adsorption, membrane separation and flotation for heavy metal removal from wastewater. *Desalination*, 2003, 157, 97-104.

Meshko, V; Markovska, L; Mincheva, M; Rodrigues, AE. Adsorption of basic dyes on granular activated carbon and natural zeolite. *Water Res.*, 2001, 35(14), 3357-3366.

Meunier, N; Laroulandie, J; Blais, JF; Tyagi, RD. Cocoa shells for heavy metal removal from acidic solutions. *Biores. Technol.*, 2003, 90(3), 255-263.

Misaelides, P; Godelitsas, A; Fillipidis, A; Charistos, D; Anousis, I. Thorium and uranium uptake by natural zeolitic materials. *Sci. Total Environ.*, 1995, 173/174, 237-246.

Mohammad, AW; Othaman, R; Hilal, N. Potential use of nanofiltration membranes in treatment of industrial wastewater from Ni-P electroless plating. *Desalination*, 2004, 168, 241-252.

Mondales, KD; Carland, RM; Aplan, FF. The comparative ion-exchange capacities of natural sedimentary and synthetic zeolites. *Min. Eng.*, 1995, 8(4-5), 535-548.

Monser, L; Adhoum, N. Modified activated carbon for the removal of copper, zinc, chromium, and cyanide from wastewater. *Sep. Purif. Technol.*, 2002, 26, 137-146.

Nakano, Y; Takeshita, K; Tsutsumi, T. Adsorption mechanism of hexavalent chromium by redox within condensed-tannin gel. *Water Res.*, 2001, 35(2), 496-500.

Navarro, RR; Sumi, K; Matsumura, M. Heavy metal sequestration properties of a new amine-type chelating adsorbent. *Water Sci. Technol.*, 1998, 38(4-5), 195-201.

Netpradit, S; Thiravetyan, P; Towprayoon, S. Application of 'waste' metal hydroxy sludge for adsorption of azo reactive dyes. *Water Res.*, 2003, 37, 763-772.

Netzer, A; Hughes, DE. Adsorption of copper, lead, and cobalt by activated carbon. *Water Res.*, 1984, 18(8), 927-933.

Ning, RY. Arsenic removal by reverse osmosis. *Desalination*, 2002, 143, 237-241.

Nourbakhsh, M; Sag, Y; Özer, D; Aksu, Z; Kutsal, T; Caglar, A. A comparative study of various biosorbents for removal of chromium(VI) ions from industrial wastewater. *Proc. Biochem.*, 1994, 29, 1-5.

Offringa, G. Dissolved air flotation in Southern Africa. Water Sci. Technol., 1995, 31(3-4), 159–172.

Orhan, G; Arslan, C; Bombach, H; Stelter, M. Nickel recovery from the rinsewaters of plating baths. *Hydrometallurgy*, 2002, 65, 1-8.

Ortiz, N; Pires, MAF; Bressiani, JC. Use of steel converter slag as nickel adsorbent to waste-water treatment. *Waste Manage.*, 2001, 21, 631-635.

Ouki, SK; Cheeseman, CR; Perry, R. Effects of conditioning and treatment of chabazite and clinoptilolite prior to lead and cadmium removal. *Environ. Sci. Technol.*, 1993, 27(6), 1108-1116.

Ouki, SK; Neufeld, RD; Perry, R. Use of activated carbon for the recovery of chromium from industrial wastewaters. *J. Chem. Technol. Biotechnol.*, 1997, 70, 3-8.

Ozaki, H; Sharma, K; Saktaywin, W. Performance of an ultra-low-pressure reverse osmosis membrane (ULPROM) for separating heavy metal: Effects of interference parameters. *Desalination*, 2002, 144, 287-294.

Panday, KK; Prasad, G; Singh, VN. Removal of Cr(VI) from aqueous solution by adsorption on fly ash-wollastonite. *J. Chem. Technol. Biotechnol.*, 1984a, 34A(7), 367-374.

Panday, KK; Prasad, G; Singh, VN. Fly ash-china clay for removal of Cr(VI) from aqueous solutions. *Ind. J. Chem.*, 1984b, 23A, 514-515.

Panday, KK; Prasad, G; Singh, VN. Copper(II) removal from aqueous solution by fly ash. *Water Res.*, 1985, 19(7), 869-873.

Pansini, M; Colella, C; De'Gennaro, M. Chromium removal from water by ion exchange using zeolite. *Desalination*, 1991, 83, 145-157.

Papadopoulos, A; Fatta, D; Parperis, K; Mentzis, A; Harambous, KJ; Loizidou, M. Nickel uptake from a wastewater stream produced in a metal finishing industry by combination of ion-exchange and precipitation methods. *Sep. Purif. Technol.*, 2004, 39(3), 181-188.

Park, JS; Jung, WY. Removal of chromium by activated carbon fibers plated with copper metal. *Carbon Sci.*, 2001, 2(1), 15-21.

Pedersen, AJ. Characterization and electrolytic treatment of wood combustion fly ash for the removal of cadmium. *Biomass Bioenergy*, 2003, 25(4), 447-458.

Peniche-Covas, C; Alvarez, LW; Arguella-Monal, W. The adsorption of mercuric ions by chitosan. *J. Appl. Polym. Sci.*, 1992, 46, 1147-1150.

Pereira, MFR, Soares, SF; Orfao, JJM; Figueiredo, JL. Adsorption of dyes on activated carbons: influence of surface chemical groups. *Carbon*, 2003, 41, 811-821.

Periasamy, K; Namasivayam, C. Removal of copper(II) by adsorption onto peanut hull carbon from water and copper plating industry wastewater. *Chemosphere*, 1996, 32(4), 769-789.

Periasamy, K; Namasivayam, C. Removal of nickel (II) from aqueous solution and nickel plating industry wastewater using an agricultural waste: peanut hulls. *Waste Manage.*, 1995, 15(1), 63-68.

Perić, J; Trgo, M; Medvidović, NV. Removal of zinc, copper, and lead by natural zeolite-a comparison of adsorption isotherms. *Water Res.*, 2004, 38(7), 1893-1899.

Persin, F ; Rumeau, M. Le traitement électrochimique des eaux et des effluents. Tribune de l'eau, 1989, 3(42), 45-46.

Pittman, CU; He, GR; Gardner, SD. Chemical modification of carbon fiber surfaces by nitric acid oxidation followed by reaction with tetraethylenepentamine. *Carbon*, 1997, 35(3), 317-331.

Pollution Control Department (PCD), Thai Ministry of Natural Resources and Environment, Water quality standards. 2004. Available from: <http://www.pcd.go.th/info_serv/en_reg_std_water04.html>

Pollution Control Department (PCD), Thai Ministry of Natural Resources and Environment, *Water quality standards.* 2005. Available from: <URL:ftp://pcd.go.th/information/Regulations/WaterQuality/Effluents. htm>

Potts, DE; Ahlert, RC; Wang, SS. A critical review of fouling of reverse osmosis membranes. *Desalination*, 1981, 36, 235-264.

Pradhan, BK; Sandle, NK. Effects of different oxidizing agent treatments on the surface properties of activated carbon. *Carbon*, 1999b, 37, 1323-1332.

Pradhan, J; Das, SN; Thakur, RS. Adsorption of hexavalent chromium from aqueous solution by using activated red mud. *J. Coll. Int. Sci.*, 1999, 217,137-141.

Qdais, HA; Moussa, H. Removal of heavy metals from wastewater by membrane processes: a comparative study. *Desalination*, 2004, 164, 105-110.

Qin, JJ; Wai, MN; Oo, MH; Wong, FS. A feasibility study on the treatment and recycling of a wastewater from metal plating. *J. Membr. Sci.*, 2002, 208, 213-221.

Rai, D; Eary, LE; Zacharia, JM. Environmental chemistry of chromium. *Sci. Total Environ.*, 1989, 86, 15-23.

Ramos, RL; Martinez, J; Coronado, RMG. Adsorption of chromium(VI) from aqueous solutions on activated carbon. *Water Sci. Technol.*, 1994, 30(9), 191-197.

Ramos, RL; Rangel-Mendez, JR; Mendoza-Barron, J; Fuentes-Rubio, L; Guerrero-Coronado, RM. Adsorption of cadmium(II) from aqueous solution onto activated carbon. *Water Sci. Tech.*, 1997, 35(7), 205-211.

Ramos, RL; Jacome, LAB; Barron, JM; Rubio, LF; Coronado, RMG. Adsorption of zinc(II) from an aqueous solution onto activated carbon. *J. Hazard. Maters.*, 2002, B90, 27-38.

Rana, P; Mohan, N; Rajagopal, C. Electrochemical removal of chromium from wastewater by using carbon aerogel electrodes. *Water Res.*, 2004, 38, 2811-2820.

Rangel-Mendez, JR; Streat, M. Adsorption of cadmium by carbon cloth: Influence of surface oxidation and solution pH. *Water Res.*, 2002, 36, 1244-1252.

Rao, M; Parawate, AV; Bhole, AG. Removal of Cr^{6+} and Ni^{2+} from aqueous solution using bagasse and fly ash. *Waste Manage.*, 2002, 22, 821-830.

Reddad, Z; Gerenete, C; Andres, Y; Le Cloeiric, P. Mechanisms of Cr(III) and Cr(VI) removal from aqueous solutions by sugar beet pulp. *Environ. Technol.*, 2003, 24, 257-264.

Reed, BE; Arunachalam, S. Use of granular activated carbon (GAC) column for lead removal. *J. Environ. Eng.*, 1994a, 120(2), 416-436.

Reed, BE; Arunachalam, S. Removal of lead and cadmium from aqueous waste stream using granular activated carbon (GAC) columns. *Environ. Prog.*, 1994b, 13(1), 60-64.

Reed, BE; Robertson, J; Jamil, M. Regeneration of granular activated carbon (GAC) column used for removing lead. *J. Environ. Eng.*, 1995, 121(9), 653-662.

Rengaraj, S; Joo, CK; Kim, YH; Yi, JH. Kinetics of removal of chromium from water and electronic process wastewater by ion exchange resins: 1200H, 1500H, and IRN97H. *J. Hazard. Mater.*, 2003, 102(2-3), 257-275.

Rengaraj, S; Yeon, KH; Moon, SH. Removal of chromium from water and wastewater by ion exchange resins. *J. Hazard. Mater.*, 2001, B87, 273-287.

Rivera-Utrilla, J; Toledo-Bautista, I; Ferro-Garcia, MA; Moreno-Castilla, C. Bioadsorption of Pb(II), Cd(II), and Cr(VI) on activated carbon from aqueous solution. *Carbon*, 2003, 41, 323-330.

Rubio, J; Tessele, F. Removal of heavy metal ions by adsorptive particulate flotation. *Min. Eng.*, 1997, 10(7), 671-679.

Sablani, SS; Goosen, MFA; Al-Belushi, R; Wilf, M. Concentration polarization in ultrafiltration and reverse osmosis: a critical review. *Desalination*, 2001, 141, 269-289.

Saffaj, N; Loukil, H; Younssi, SA; Albizane, A; Bouhria, M; Persin, M; Larbot, A. Filtration of solution containing heavy metals and dyes by means of ultrafiltration membranes deposited on support made of Morrocan clay. *Desalination*, 2004, 168, 301-306.

Sapari, N; Idris, A; Hisham, N. Total removal of heavy metal from mixed plating rinse wastewater. *Desalination*, 1996, 106, 419-422.

Schmuhl, R; Krieg, HM; Keizer, K. Adsorption of Cu(II) and Cr(VI) ions by chitosan: kinetics and equilibrium studies. *Water SA.*, 2001, 27(1), 1-7.

Selvakumari, G; Murugesan, M; Pattabi, S; Sathishkumar, M. Treatment of electroplating industry effluent using maize cob carbon. *Bull. Environ. Contam. Toxicol.*, 2002, 69, 195-202.

Selvaraj, K; Manonmani, S; Pattabhi, S. Removal of hexavalent chromium using distillery sludge. *Biores.Technol.*, 2003, 89, 207-211.

Selvi, K; Pattabi, S; Kadirvelu, K. Removal of Cr(VI) from aqueous solution by adsorption onto activated carbon. *Biores. Technol.*, 2001, 80, 87-89.

Semerjian, L; Ayoub, GM. High-pH-magnesium coagulation-flocculation in wastewater treatment. *Adv. Environ. Res.*, 2003, 7, 389-403.

Semmens, MJ; Martin, WP. The influence of pretreatment on the capacity and selectivity of clinoptilolite for metal removal. *Water Res.*, 1988, 22(5), 537-542.

Seron, A; Bennadi, H; Beguin, F; Frackowiak, E; Bretelle, JL; Thity, MC; Bandoz, TJ; Jagiello, J; Schwarz, JA. Sorption and desorption of lithium ions from activated carbons. *Carbon*, 1996, 34(4), 481-487.

Shammas, NK. Coagulation and flocculation. In: *Physicochemical Treatment Processes*; Wang, LK; Hung, YT; Shammas NK.; Eds.; Humana Press: New Jersey, 2004; Vol. 3, pp 103-140.

Sharma, DC; Forster, CF. Removal of hexavalent chromium using sphagnum moss peat. *Water Res.*, 1993, 27(7), 1201-1208.

Sharma, DC; Forster, CF. Continuous adsorption and desorption of chromium ions by sphagnum moss peat. *Process Biochem.*, 1994a, 30(4), 293-298.

Sharma, DC; Forster, C.F. A preliminary examination into the adsorption of hexavalent chromium using low cost adsorbents. *Biores. Technol.*, 1994b, 47, 257-264.

Sharma, DC; Forster, CF. Column studies into the adsorption of chromium(VI) using sphagnum moss peat. *Biores. Technol.*, 1995, 52, 261-267.

Sharma, DC; Forster, CF. Removal of hexavalent chromium from aqueous solutions by granular activated carbon. *Water SA.*, 1996a, 22, 153-160.

Sharma, DC; Forster, CF. A comparison of the sorptive characteristics of leaf mould and activated carbon columns for the removal of hexavalent chromium. *Process Biochem.*, 1996b, 31(3), 213-218.

Sharma, YC. Cr(VI) removal from industrial effluents by adsorption on an indigenous low-cost material. *Coll. Surf. A.*, 2003, 215(1-3), 155-162.

Shawabkeh, RA; Rockstraw, DA; Bhada, RK. Copper and strontium adsorption by a novel carbon material manufactured from pecan shells. *Carbon*, 2002, 40, 781-786.

Shim, JW; Lee, SM; Rhee, BS; Ryu, SK. Adsorption of Ni(II), Cu(II), Cr(VI) from multi-component aqueous solution by pitch-based ACF, In: *Proceeding of The European Carbon Conference "Carbon 96"*, New Castle, UK, 1996.

Shim, JW; Park, SJ; Ryu, SK. Effect of modification with HNO_3 and NaOH by pitch-based activated carbon fibers. *Carbon*, 2001, 39, 1635-1642.

Shukla, A; Zhang, YH; Dubey, P; Margrave, JL; Shukla, SS. The role of sawdust in the removal of unwanted materials from water. *J. Hazard. Maters.*, 2002, 2884, 1-16.

Shukla, SR; Pai, RS. Comparison of Pb(II) uptake by coir and dye loaded coir fibres in a fixed bed column. *J. Hazard. Mater.*, 2005, B125, 147-153.

Singh, AK; Singh, DP; Singh, VN. Removal of Zn(II) from water by adsorption on china clay. *Environ. Technol. Lettrs.* 1998, 9, 1153-1162.

Singh, IB; Singh, R. Effects of pH on Cr-Fe interaction during Cr(VI) removal by metallic iron. *Environ. Technol.*, 2003, 24, 1041-1047.

Slater, CS; Ahlert, RC; Uchrin, CG. Applications of reverse osmosis to complex industrial wastewater treatment. *Desalination*, 1983, 48, 171-187.

Smith, EH; Lu, WP; Vengris, T; Binkiene, R. Sorption of heavy metals by Lithuanian glauconite. *Water Res.*, 1996, 30(12), 2883-2892.

Solé, M; Casas, JM; Lao, C. Removal of Zn from aqueous solutions by low-rank coal. *Water, Air, Soil Pollut.*, 2003, 144, 57-65.

Standard Methods for the Examination of Water and Wastewater 20th. Washington: American Public Health Association (APHA); 1998.

Srinivasan, K; Balasubramanian, N; Ramakhrisna, TV. Studies on chromium removal by rice husk carbon. *Ind. J. Environ. Health.*, 1988, 30(4), 376-387.

Srivastava, SK; Bhattacharjee, G; Tyagi, R; Pant, N; Pal, N. Studies on the removal of some toxic metal ions from aqueous solutions and industrial waste part I: removal of lead and cadmium by hydrous iron and aluminium oxide. *Environ. Technol. Lettrs.*, 1989a, 9, 1173-1185.

Srivastava, SK; Tyagi, R; Pal, N. Studies on the removal of some toxic metal ions part II: removal of lead and cadmium by montmorillonite and kaolinite. *Environ. Technol. Lettrs.*, 1989b, 10, 275-282.

Srivastava, SK; Tyagi, R; Pant, N. Adsorption of heavy metal ions on carbonaceous material developed from the waste slurry generated in local fertilizer plants. *Water Res.*, 1989c, 23(9), 1161-1165.

Srivastava, SK; Gupta, VK; Mohan, D. Removal of lead and chromium by activated slag-a blast-furnace waste. *J. Environ. Eng.*, 1997, 123(5), 461-468.

Strelko, V; Malik, DJ; Streat, M. Characterisation of the surface of oxidized carbon adsorbents. *Carbon*, 2002, 40, 95-104.

Subbaiah, T; Mallick, SC; Mishra, KG; Sanjay, K. Das, R.P. Electrochemical precipitation of nickel hydroxide. *J. Power Sources*, 2002, 112, 562-569.

Tan, WT; Ooi, ST; Lee, C.K. Removal of chromium(VI) from solution by coconut husk and palm pressed fibers. *Environ. Technol.*, 1993, 14, 277-282.

Tandon, RK; Crisp, PT; Ellis, J. Effect of pH on chromium(VI) species in solution. *Talanta,* 1984, 31(3), 227-228.

Tang, PL; Lee, CK; Low, KS; Zainal, Z. Sorption of Cr(VI) and Cu(II) in aqueous solution by ethylenediamine modified rice hull. *Environ. Technol.*, 2003, 24, 1243-1251.

Tavares, CR; Vieira, M; Petrus, JCC; Bortoletto, EC; Ceravollo, F. Ultrafiltration/ complexation process for metal removal from pulp and paper industry wastewater. *Desalination*, 2002, 144, 261-265.

Tsagarakis, KP; Mara, DD; Angelakis, AN. Application of cost criteria for selection of municipal wastewater treatment systems. *Water, Air, Soil Pollut.*, 2003, 142, 187-210.

Tünay, O. Developments in the application of chemical technologies to wastewater treatment. *Water Sci. Technol.*, 2003, 48(11-12), 43-52.

Tünay, O. Kabdasli, NI. Tasli, R. Pretreatment of complexed metal wastewaters. *Water Sci. Technol.*, 1994, 29(9), 265-274.

Tünay, O; Kabdasli, N.I. Hydroxide precipitation of complexed metals. *Water Res.*, 1994, 28(10), 2117-2124.

Tzanetakis, N; Taama, WM; Scott, K; Jachuck, RJJ; Slade, RS; Varcoe, J. Comparative performance of ion exchange membrane for electrodialysis of nickel and cobalt. *Sep. Purif. Technol.*, 2003, 30, 113-127.

Udaybhaskar, P; Iyengar, L; Prabhakara Rao, AVS. Hexavalent chromium interaction with chitosan. *J. Appl. Polym. Sci.*, 1990, 39, 739-747.

Ujang, Z; Anderson, GK. Application of low-pressure reverse osmosis membrane for Zn^{2+} and Cu^{2+} removal from wastewater. *Water Sci. Technol.*, 1996, 34(9), 247-253.

Ulmanu, M; Maranon, E; Fernandez, Y; Castrillon, L; Anger, I; Dumitriu, D. Removal of copper and cadmium ions from diluted aqueous solutions by low cost and waste material adsorbents. *Water, Air, Soil Pollut.*, 2002, 142(1-4), 357-373.

Vaca-Mier, M; Callejas, RL; Gehr, R; Cisneros, BEJ; Alvarez, PJJ. Heavy metal removal with Mexican clinoptilolite: multi-component ionic exchange. *Water Res.*, 2001, 35(2), 373-378.

Vengris, T; Binkiene, R; Sveikauskaite, A. Nickel, copper, and zinc removal from wastewater by a modified clay sorbent. *Appl. Clay Sci.*, 2001, 18, 183-190.

Vigneswaran, S; Ngo, HH; Chaudhary, DS; Hung, YT. Physico-chemical treatment processes for water reuse. In: *Physicochemical Treatment Processes*; Wang, LK; Hung, YT; Shammas, NK.; Eds.; Humana Press: New Jersey, 2004; Vol. 3, pp 635-676.

Vik, EI; Carlsoon, DA; Eikum, AS; Gjessing, ET. Electrocoagulation of potable water. *Water Res.*, 1984, 18, 1355-1360.

Virta, R. USGS mineral information, US geological survey mineral commodity summary 2000. 2001. Available from: <http://www.minerals.usgs.gov/minerals/pubs/commodity/zeolite/zeomyb00.pdf>

Virta, R. USGS mineral information, US geological survey mineral commodity summary 2002. 2002. Available from: <http://www.minerals.usgs.gov/minerals/pubs/commodity/clays/190496.pdf>

Vrijenhoek, EM; Waypa, JJ. Arsenic removal from drinking water by a "loose" nanofiltration membrane. *Desalination*, 200, 130, 265-277.

Wang, LK; Fahey, EM; Wu, ZC. Dissolved air flotation. In: *Physicochemical Treatment Processes*; Wang, LK; Hung, YT; Shammas, NK.; Eds.; Humana Press: New Jersey, 2004; Vol. 3, pp 431-500.

Wang, LK; Vaccari, DA; Li, Y; Shammas, NK. Chemical precipitation In: *Physicochemical Treatment Processes*; Wang, LK; Hung, YT; Shammas NK.; Eds.; Humana Press: New Jersey, 2004; Vol. 3, pp141-198.

Weber, B; Holz, F. Combination of activated sludge pre-treatment and reverse osmosis. In: *Landfilling of Waste Leachate*; Christensen, TH; Cossu, R; Stegmann, R.; Eds.; Elsevier: Amsterdam, 1992; pp. 323-332.

Weng, CH. Removal of heavy metals by fly ash. University of Delaware, Delaware; 1990. (Master thesis)

Weng, CH. Removal of nickel(II) from dilute aqueous solution by sludge-ash. *J. Environ. Eng.*, 2002, 128(8), 716-721.

Weng, CH; Wang, JH; Huang, CP. Adsorption of Cr(VI) onto TiO_2 from dilute aqueous solutions. *Water Sci. Technol.*, 1997, 35(7), 55-62.

Wingenfelder, U; Hansen, C; Furrer, G; Schulin, R. Removal of heavy metals from mine water by natural zeolites. *Environ. Sci. Technol.*, 2005, 39, 4606-4613.

Wingenfelder, U; Nowack, B; Furrer, G; Schulin, R. Adsorption of Pb and Cd by amine-modified zeolite. *Water Res.*, 2005, 39, 3287-3297.

Wong, FS; Qin, JJ; Wai, MN; Lim, AL; Adiga, M. A pilot study on a membrane process for the treatment and recycling of spent final rinse water from electroless plating. *Sep. Purif. Technol.*, 2002, 29(1), 41-51.

Wong, KK; Lee, CK; Low, KS; Haron, MJ. Removal of Cu and Pb by tartaric acid modified rice husk from aqueous solutions. *Chemosphere*, 2003, 50, 23-28.

Yadava, KP; Tyagi, BS; Singh, VN. Effect of temperature on the removal of lead(II) by adsorption on china clay and wollastonite. *J. Chem. Biotechnol.*, 1991, 51, 47-60.

Yang, XJ; Fane, AG; MacNaughton, S. Removal and recovery of heavy metals from wastewater by supported liquid membranes. *Water Sci. Technol.*, 2001, 43(2), 341-348.

Yavuz, O; Altunkaynak, Y; Guzel, F. Removal of copper, nickel, cobalt, and manganese from aqueos solution by kaolinite. *Water Res.*, 2003, 37, 948-952.

Yu, LJ; Shukla, SS; Dorris, KL; Shukla, A; Margrave, JL. Adsorption of chromium from aqueous solutions by maple sawdust. *J. Hazard. Mater.*, 2003, 100(1-3), 53-63.

Yurlova, L; Kryvoruchko, A; Kornilovich, B. Removal of Ni(II) ions from wastewater by micellar-enhanced ultrafiltration. *Desalination*, 2002, 144, 255-260.

Zabel, T. Flotation in water treatment. In: *The Scientific Basis of Flotation*; Ives, KJ.; Ed.; Martinus Nijhoff Publishers; Hague, 1984; pp 349-378.

Zamboulis, D; Pataroudi, SI; Zouboulis, AI; Matis, KA. The application of sorptive flotation for the removal of metal ions. *Desalination*, 2004, 162, 159-168.

Zamzow, MJ; Eichbaum, BR; Sandgren, KR; Shanks, DE. Removal of heavy metal and other cations from wastewater using zeolites. *Sep. Sci. Technol.*, 1990, 25(13-15), 1555-1569.

Zhai, Y; Wei, XX; Zeng, GM; Zhang, D; Chu, KF. Study of adsorbent derived from sewage sludge for the removal of Cd^{2+}, Ni^{2+} in aqueous solution. *Sep. Purif. Technol.*, 2004, 38, 191-196.

Zouboulis, AI; Kydros, KA. Use of red mud for toxic metals removal: the case of nickel. *J. Chem. Technol. Biotechnol.*, 1993, 58, 95-101.

Zouboulis, AI; Matis, KA; Lazaridis, NK; Golyshin, PN. The use of biosurfactants in flotation: application for the removal of metal ions. *Min. Eng.*, 2003, 16, 1231-1236.

Zoubulis, AI; Kydros, KA; Matis, KA. Removal of hexavalent chromium anions from solutions by pyrite fines. *Water Res.*, 1995, 29(7), 1755-1760.

APPENDICES

APPENDIX 1. CURVE CALIBRATION OF STANDARD HEXAVALENT CHROMIUM SOLUTION

Volume of Cr standard solution (20 mg/L) taken (mL)	Amount of deionized water added (mL)	Chromium Concentration (mg)	Absorbance at $\lambda = 540$ nm
0.25	9.75	0.50	0.437
0.50	9.50	1.00	0.777
0.75	9.25	1.50	1.118
1.00	9.00	2.00	1.459
1.25	8.75	2.50	1.799
1.50	8.50	3.00	2.140

Calibration Curve: $Y= 0.0961 + 0.6813X$, $R^2= 0.9999$ where Y= Absorbance at wavelength (λ=540 nm and X= Concentration of Standard Hexavalent Chromium Solution (mg/L).

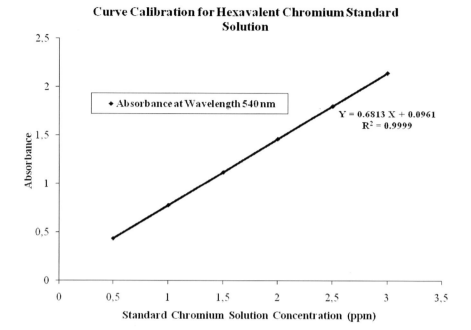

Curve Calibration for Hexavalent Chromium Standard Solution

APPENDIX 2

BioInfoBank
L i b r a r y

Diploma

The Scientific Committee of BioInfoBank Institute has
selected a MSc Thesis entitled

*A Research Study on Cr(VI) Removal from Electroplating
Wastewater Using Chemically Modified Low-cost Adsorbents
and Commercial Activated Carbon.*

written by

Tonni Agustiono Kurniawan

as the BEST THESIS submitted to the
BioInfoBank Library in 2008.

Leszek Rychlewski
Scientific Director of the BioInfoBank Institute

APPENDIX 3

0452

(Official Translation)

Thammasat University

The University Council hereby proclaims that

Mr.Tonni Agustiono Kurniawan

has successfully fulfilled the degree requirements in the academic year 2002

Thereby the University has conferred upon him the degree of

Master of Science in Technology

May the recipient be blessed with happiness and prosperity
Granted with the University Council's approval
on the 14th day of May, 2003

(Panas Simasatien)
Chairperson, the University Council

(Naris Chaiyasoot)
Rector

(Prida Wibulswas)
Director, Sirindhorn International Institute of Technology

Certified by:

S. Tantarat

Deputy Director

W. Jeonkde

Registrar

Sirindhorn International Institute of Technology

APPENDIX 4

November 29, 2011

To Whom It May Concern:

This letter is to inform you that the work of Dr. Tonni Agustiono Kurniawan currently ranks among the top 1% of researchers in the field of Engineering in *Essential Science Indicators*[SM] from Thomson Reuters.

Essential Science Indicators is a compilation of science indicators and trend data derived from the Thomson Reuters databases, focusing on highly cited papers, scientists, organizations, journals, and nations. It combines these data with editorial content to highlight important results. What makes *Essential Science Indicators* a unique tool is that it is completely and exclusively Web-based, including the editorial features, which are available to the Web community at large via ScienceWatch.com (http://sciencewatch.com).

Dr. Kurniawan's record in this field as of November 2011 includes 4 papers cited a total of 651 times between January 1, 2001 and August 31, 2011. Dr. Kurniawan's most-cited paper in this field is "Low-cost adsorbents for heavy metals uptake from contaminated water: a review," (Babel S, Kurniawan TA, *J. Hazard. Mater.* 97: 219-43, 28 February 2003). This paper was initially chosen in December 2004 as a Fast-Breaking Paper in Engineering. At that time, the paper had attracted 13 cites. Dr. Kurniawan and Dr. Babel graciously granted us an interview about this paper, and it is archived at: http://www.esi-topics.com/fbp/2004/december04-SandhyaBabel.html. Its current citation count in our database is 543 cites.

If you have any questions, comments, or concerns, please contact me at Jennifer.Minnick@thomsonreuters.com.

Sincerely,

Jennifer L. Minnick
Editor, *Essential Science Indicators* and *ScienceWatch.com*

ABOUT THE AUTHOR

Presently Dr. Tonni Agustiono Kurniawan is a Green Talent Researcher at the Ravensburg-Weingarten University of Applied Sciences (Germany) with the support of the German Federal Ministry for Education and Research (BMBF). As one of the 2010 Green Talents, his credential as an outstanding early-career researcher in the field of environmental technology has been internationally recognized by his peers through an exceptional citation counts of his publication entitled 'Low cost adsorbents for heavy metals uptake from contaminated water: a review' in the Journal of Hazardous Materials B97:219-243 (2003). Based on the citation database of the Scopus and Web of Science, the article has topped the journal's list of highly-cited articles since its establishment in 1975. The research work, presented in this book, won an international recognition from the BioInfo Bank Institute (Poland) in 2008 as the Best Thesis submitted across disciplines.

As soon as he completed his Ph.D. study at The Hong Kong Polytechnic University in 2008, Dr. Kurniawan had undertaken his postdoctoral research at the Kuopio University (Finland) and the United Nations University-Institute of Advanced Studies (Japan). To date, with citation of over 1300 counts and an h-index of 11, he has authored and/or co-authored over 50 scientific publications in various SCI journals with high impact factor, conference proceedings, monographs and book chapters. Recently the Essential Science Indicators of Institute for Scientific Information (ISI)-Thomson Reuter identified him as one of most highly cited top 1% scientist in the field of engineering.

INDEX

A

access, vii
acetic acid, 84, 94
acetone, 85
ACF, 39, 40, 180
acid, ix, 7, 8, 14, 22, 23, 28, 30, 31, 35, 37,
    39, 40, 41, 45, 83, 84, 93, 94, 95, 125,
    129, 130, 134, 135, 136, 137, 138, 151,
    152, 153, 157, 164, 169, 171, 174, 177,
    183
acidic, 9, 18, 22, 23, 24, 27, 30, 31, 37, 38,
    40, 89, 91, 85, 89, 92, 94, 95, 96, 100,
    129, 131, 144, 151, 160, 175
activated carbon, ix, xv, 1, 5, 6, 7, 9, 13, 17,
    27, 31, 35, 39, 40, 41, 42, 45, 81, 93, 81,
    125, 154, 155, 163, 164, 165, 166, 167,
    168, 169, 170, 171, 172, 173, 174, 175,
    176, 177, 178, 179, 180
adjustment, 13, 14, 15, 87
adsorption isotherms, 177
adverse effects, 20
Africa, 175
agencies, 155
aggregation, 4, 15, 16
alternative treatments, 27
aluminium, 180
amine, 175, 183
amino, 18, 19
amino groups, 18
ammonium, 17, 89, 160
ANOVA, xv, 9, 88, 130, 134

aqueous solutions, 8, 165, 167, 169, 171,
    172, 174, 176, 177, 178, 179, 180, 182,
    183
arsenic, 166, 167, 168
Asia, 44
assessment, 44
ataxia, 12
atoms, 19
attachment, 16

B

base, 95, 180
baths, 26, 176
benefits, 22, 43
biodegradability, 81
biodegradation, 20
biological processes, 156
biomass, 171
bonding, 17
bonds, 96
by-products, 27, 28, 32, 33

C

Ca^{2+}, 127
cadmium, 6, 160, 163, 164, 165, 170, 172,
    173, 176, 177, 178, 180, 182
calcium, 14, 127
carbohydrate, 30
carbon dioxide, 28

carbonyl groups, 38
carboxyl, 31, 96
carboxylic groups, 30, 95
catalyst, 89
cation, 7, 14, 18, 23, 25, 32, 34
CEC, 7, 32, 34
cellulose, 6, 18, 30, 41, 174
cesium, 7
charge density, 35
chelates, 160
chemical characteristics, ix, 81, 153, 155
chemical interaction, 27, 88, 92, 134, 142, 159, 161
chemical properties, 82
chemical stability, 22
chemicals, 5, 8, 15, 27, 29, 31, 42, 43, 81, 91, 87, 164
children, vii
China, 12
chitosan, ix, xv, 6, 7, 8, 9, 18, 19, 28, 83, 84, 93, 81, 93, 136, 137, 150, 153, 154, 156, 157, 158, 165, 169, 171, 174, 176, 179, 181
Chitosan, 82, 83, 84, 165
chlorine, 22
City, 85
clay minerals, 36
closure, 171
coagulation process, 15
coal, 6, 36, 37, 180
cobalt, 6, 175, 181, 183
cocoa, 28
coconut shell charcoal (CSC), ix, 1, 6, 11, 28, 93, 155
coffee, 38, 166
color, 91
combustion, 176
commercial, ix, 1, 5, 7, 9, 24, 25, 33, 37, 41, 44, 81, 83, 93, 81, 125, 152, 154, 155, 165, 172
commercial activated carbon, ix, 1, 5, 9, 81, 93, 81, 125, 154, 155, 165, 172
commodity, 182
competition, 96, 169
complement, 127

composition, 35, 43
compounds, 3, 14, 18, 21, 22, 32, 82, 95, 159
concentration of chromium, ix, 87
conditioning, 83, 176
conductivity, 82, 96, 165
conference, xiii
CONGRESS, 165
constituents, 6
construction, 43
consumption, 13, 16, 22, 26, 27, 30, 31, 43, 45, 94, 152
contact time, 8, 24, 31, 87, 88, 89, 92, 83, 89, 101, 129, 134, 135, 142, 153, 156, 157, 159
contaminant, 13
contaminated water, vii, xiii, 1, 2, 3, 4, 6, 7, 11, 28, 87, 97, 130, 133, 136, 144, 149, 155, 156, 159, 165
contamination, 81
COOH, 38, 92, 94
cooling, 84
coordination, 18
copper, 15, 160, 164, 165, 166, 167, 169, 170, 171, 172, 173, 175, 176, 177, 182, 183
Copyright, 9, 93, 154
correlation, 135
correlation coefficient, 135
correlations, 89, 92, 157
cost effectiveness, 42, 152
cost minimization, 41
cost-benefit analysis, 42, 161
coughing, 12
crystalline, 7
customers, 2
cyanide, 166, 175
cycles, 140, 142, 149, 158, 159

D

damages, 25
data set, 9, 92
database, xiii
decision makers, vii

decision-making process, 42
decontamination, 11
deposition, 168
depth, 42, 159, 161
desorption, 28, 31, 139, 142, 144, 146, 148,
    158, 159, 179
detectable, 92
detention, 4
developed countries, 4
developing countries, vii, 3
deviation, 9, 92
diarrhea, 3, 12
diffusion, 85, 87, 89
diffusion process, 89
disinfection, 173
disorder, 12
dissolved oxygen, 89
distribution, 92, 160, 161
donors, 30, 94
drainage, 168, 169
drinking water, vii, 166, 172, 182
dyes, 164, 175, 176, 178

energy consumption, 13, 22, 26, 27, 43, 45
engineering, vii, xiii, 2, 44, 85
England, 82, 85
environment, vii, 1, 3, 4, 11, 12, 16, 45, 81,
    153, 155, 172
environmental conditions, 41
environmental impact, 15, 45
environmental protection, vii, 13, 171
Environmental Protection Agency, 2, 168
environmental regulations, 3, 5, 44
environmental technology, xiii, 3
environments, 6, 12
EPA, 2, 3, 8, 13, 14, 81, 157, 158, 168
equilibrium, 2, 23, 28, 29, 30, 32, 33, 34,
    35, 36, 37, 38, 87, 88, 89, 83, 89, 90, 91,
    96, 101, 134, 135, 144, 168, 172, 174,
    179
equilibrium sorption, 23, 28, 29, 30
equipment, 15, 24, 43, 85, 86
Europe, 4
European Community, 36
exclusion, 21
exposure, 127, 146

E

economics, 31, 45
effluents, 1, 3, 4, 5, 8, 11, 88, 89, 90, 81,
    152, 155, 156, 158, 161, 168, 169, 172,
    177, 180
electrical conductivity, 96
electricity, 8, 42
electrochemistry, 171
electrodes, 25, 26, 27, 178
electroflotation, 16
electrolysis, 13, 25
electron, 18, 30, 94, 95, 96
electrons, 95, 96
electroplating, 1, 2, 3, 4, 5, 8, 11, 12, 15, 20,
    26, 44, 81, 82, 86, 89, 93, 81, 82, 84,
    152, 153, 155, 159, 160, 163, 164, 165,
    169, 170, 172, 173, 174, 179
employees, 2
employment, 2
endothermic, 32, 33, 36, 160
energy, 13, 22, 26, 27, 43, 45, 90

F

FAS, 89
fiber, 40, 177
fibers, 6, 176, 180, 181
filters, 84, 87
filtration, 11, 12, 13, 17, 18, 21, 22, 44, 166,
    171, 173
Finland, xiii
flexibility, 43, 45
flocculation, 3, 11, 12, 13, 15, 16, 44, 173,
    179
flotation, 11, 12, 13, 16, 17, 43, 44, 166,
    168, 171, 173, 174, 175, 178, 182, 183,
    184
food, 3, 12
food chain, 3, 12
force, 20, 87
formaldehyde, 29
formation, 18, 19, 32, 91, 95, 131, 134, 136
formula, 89

fouling, 20, 22, 167, 177
Freundlich model, ix

G

gel, 84, 169, 175
Germany, xi, xiii, 85
Greece, 12, 34
groundwater, 166, 170
growth, 2
guidelines, 168

H

harmful effects, vii
health, 1, 3, 12
heavy metals, xiii, 1, 4, 5, 6, 11, 12, 13, 14,
    15, 16, 18, 19, 22, 23, 24, 25, 33, 40, 44,
    45, 163, 164, 165, 166, 167, 168, 171,
    173, 177, 178, 180, 182, 183
hemicellulose, 6, 41
HM, 179
Hong Kong, xiii, 168
human, 3, 12
human body, 3, 12
hydrogen, 38, 94, 95, 160
hydrogen peroxide, 160
hydrolysis, 38, 94, 95, 98, 128
hydrophilicity, 21
hydroxide, 4, 6, 8, 14, 15, 29, 164, 181
hydroxyl, 6, 30, 31, 37, 38, 41, 92, 94, 95
hydroxyl groups, 30, 94, 95

I

ideal, 13
impurities, 16, 25, 43, 83, 82, 84, 87, 97,
    100, 127, 130, 136, 139
incubator, 83, 84
India, 27
industrial processing, 32
industries, 1, 2, 3, 4, 6, 11, 12, 27, 42, 44,
    81, 153, 159

industry, 2, 4, 8, 12, 21, 28, 32, 33, 44, 82,
    89, 93, 81, 82, 84, 153, 155, 166, 168,
    170, 171, 176, 177, 179, 181
ineffectiveness, 138
inefficiency, 153
initial chromium concentration, ix, 136
insomnia, 12
interface, 24, 31, 101, 131, 142, 159
interference, 36, 176
investment, 4, 45
ion exchangers, 172
ion-exchange, 24, 167, 169, 175, 176
ionization, 87
iron, 6, 32, 33, 164, 168, 173, 180
isotherms, ix, 8, 34, 89, 92, 125, 127, 137,
    177
issues, 41
Italy, 13, 27

J

Japan, xi, xiii, 4, 173

K

K^+, 127
kinetics, ix, 4, 8, 23, 30, 139, 170, 171, 172,
    179
Korea, 13

L

laboratory studies, 1, 7, 81
lead, 163, 164, 165, 167, 169, 171, 175,
    176, 177, 178, 180, 181, 183
legislation, 1, 8, 11, 21, 22, 81, 155, 156,
    158, 168, 169
lethargy, 12
ligand, 14, 19
lignin, 6, 30, 41
liquid phase, 5, 16, 22, 27
liquids, 16
lithium, 179
local conditions, 13, 44

localization, 96
love, xi
lying, 127

M

macromolecules, 18
magnesium, 127, 179
Malaysia, 21
management, 169
manganese, 183
manpower, 43
mass, 5, 27, 31, 42, 90, 87, 89, 101, 142,
    146, 150, 161
mass loss, 31
materials, 5, 17, 20, 24, 27, 28, 34, 36, 37,
    39, 40, 41, 90, 81, 175, 180
matter, 89, 84, 169
MCP, 167
media, 86, 168, 173
membranes, 19, 20, 21, 24, 25, 174, 175,
    177, 179, 183
mercury, 165
metal ion, 4, 6, 7, 12, 13, 14, 16, 19, 23, 26,
    38, 41, 86, 88, 84, 90, 91, 94, 95, 160,
    164, 165, 166, 169, 171, 172, 174, 178,
    180, 181, 183, 184
metal ions, 4, 6, 7, 12, 13, 14, 16, 19, 23,
    26, 38, 41, 86, 88, 84, 90, 91, 94, 95,
    160, 164, 165, 166, 169, 171, 172, 174,
    178, 180, 181, 183, 184
metal recovery, 4, 5
metals, xiii, 1, 4, 5, 6, 7, 11, 12, 13, 14, 15,
    16, 17, 18, 19, 22, 23, 24, 25, 27, 32, 33,
    35, 36, 38, 40, 42, 44, 45, 127, 163, 164,
    165, 166, 167, 168, 171, 173, 174, 177,
    178, 180, 181, 182, 183, 184
meter, 82, 85
Mg^{2+}, 127
microemulsion, 167
modifications, ix, 7, 8, 19, 83, 85, 152, 158,
    160
mole, 15
molecular weight, 18

monolayer, 28, 29, 32, 33, 34, 35, 37, 38,
    90, 91, 126, 127
Moon, 178

N

Na^+, 9, 127, 128, 144, 152, 159
NaCl, ix, 34, 41, 84, 96, 97, 98, 99, 100,
    101, 125, 126, 127, 129, 137, 138, 143,
    144, 145, 146, 151, 156, 157, 158
natural resources, 7
nausea, 3, 12
neutral, 20
nickel, 160, 163, 164, 165, 170, 172, 174,
    176, 177, 181, 182, 183, 184
Nigeria, 27
nitrates, 166
nitrogen, 18

O

OH, 14, 17, 36, 87, 95, 96, 128, 144
operations, ix, 5, 7, 15, 35, 36, 85, 86, 90,
    81, 139, 142, 147, 152, 161, 170
ores, 36
organic compounds, 18, 82
organic matter, 89, 84, 169
organism, 3
osmosis, 4, 13, 18, 20, 21, 43, 45, 166, 170,
    175, 176, 177, 178, 180, 181, 182
osmotic pressure, 21
ox, 95
oxidation, 23, 25, 26, 27, 38, 89, 84, 94, 95,
    96, 160, 168, 177, 178
oxygen, xv, 34, 89, 83, 84, 95, 96, 151, 160
ozone, 160

P

pain, 3
peat, 6, 37, 38, 168, 179
permeable membrane, 18
permission, 9, 45, 93, 154
peroxide, 160

phosphate, 164
physical characteristics, 83, 158
physical properties, 5
physicochemical characteristics, 39
pilot study, 183
pitch, 180
pith, 6, 172, 174
plants, 32, 33, 34, 43, 44, 174, 181
playing, 2
Poland, vii, xiii
polarization, 178
policy, vii
pollutants, vii, 2, 166, 169
pollution, 1, 3, 11, 155, 159
polyacrylamide, 15
polymer, 38
polyvinyl alcohol, 20
poor performance, 138
population, vii
porosity, 21, 161
porous media, 168
potassium, 92, 127
power plants, 33
prayer, xi
precipitation, 3, 4, 11, 12, 13, 14, 15, 16, 22,
    24, 26, 27, 32, 43, 44, 164, 165, 167,
    168, 172, 176, 181, 182
pretreatment of zeolite, ix, 144, 152, 157,
    158
profit, 2
project, 1, 11, 155
protection, vii, 13, 171
protons, 95
public concern, 3
public health, 1
pulp, 6, 28, 178, 181
purification, 6, 16, 27, 34, 43, 169
PVP, 38, 169
pyrite, 36, 184

Q

quality standards, 177

R

radioactive waste, 7
radius, 36
raw materials, 39
RE, xv, 91, 142, 146, 150, 164
REA, xv, 87
reactions, 25, 26, 91, 95
reactivity, 5, 27, 29
reagents, 22, 81
recognition, xiii
recommendations, 155, 159
recovery, 4, 5, 16, 23, 25, 26, 34, 139, 142,
    146, 148, 163, 171, 172, 174, 176, 183
recovery process, 23, 171
recycling, 177, 183
red mud, 6, 32, 33, 177, 184
regenerate, 8, 28, 144, 146
regeneration, 8, 28, 31, 33, 34, 37, 42, 91,
    139, 140, 142, 146, 148, 149, 150, 158,
    161, 169, 174
regulations, 3, 5, 44
rejection, 18, 19, 20, 21, 25, 43, 45, 166
reliability, 9, 45, 92
requirements, 1, 5, 11, 81, 158
researchers, vii, 14, 19, 20, 23
residues, 166
resins, 4, 23, 24, 29, 169, 171, 178
resistance, 22, 87
resources, 5, 6, 7, 28
reusability, 91
reverse osmosis, 4, 13, 18, 20, 21, 43, 45,
    166, 170, 175, 176, 177, 178, 180, 181,
    182
rice husk, 6, 29, 163, 166, 180, 183
room temperature, 84
roughness, 21
rubber, 170

S

salts, 15, 20
saturation, 90
sawdust, 6, 27, 170, 180, 183
scaling, 22

scholarship, xi
science, xv, 11
scientific publications, xiii
scope, 1
sedimentation, 15
sediments, 36
selectivity, 127, 179
sewage, 37, 183
shape, 39, 161
signs, 12
silica, 30, 169
Singapore, 85
skin, 20, 21
slag, 32, 33, 168, 170, 176, 181
sludge, 4, 5, 6, 15, 16, 24, 28, 37, 42, 43, 146, 152, 156, 166, 175, 179, 182, 183
society, vii
sodium, 8, 9, 15, 17, 19, 35, 83, 94, 127, 152
sodium dodecyl sulfate (SDS), 19, 35
sodium hydroxide, 8
solid phase, 13
solid waste, vii, 4, 31, 33, 171
solubility, 3, 12, 14
sorption, 6, 23, 28, 29, 30, 32, 36, 37, 41, 90, 134, 135, 149, 166, 168, 169, 170, 174
sorption process, 6, 41
Southeast Asia, vii, 44
SP, 85
Spain, 13, 21
speciation, 24, 87, 172
species, 2, 24, 35, 36, 38, 87, 89, 95, 98, 161, 181
spending, 42
SS, 173, 174, 177, 178, 180, 183
stability, 16, 22
standard deviation, 9, 92
state, 4, 23, 26, 38, 91, 94, 160
statistics, 6
steel, 32, 33, 176
steel industry, 32
stock, 82, 87, 90
storage, 82
stormwater, 173

strong interaction, 20
strontium, 7, 180
structure, 5, 8, 35, 96, 148, 168
sugar beet, 6, 28, 178
sugar industry, 33, 170
sulfate, 17, 19, 35, 89
sulfur, 174
sulfur dioxide, 174
sulfuric acid, 7, 28, 85, 93, 125, 130, 134, 136, 137, 138, 139, 153, 157, 160
supervision, xi
supervisors, xi
supported liquid membrane, 183
surface area, 5, 23, 25, 27, 31, 41, 83, 96, 127, 135
surface chemistry, 164
surface modification, ix, 7, 28, 29, 30, 39, 45, 81, 85, 92, 81, 83, 85, 129, 138, 152, 153, 158
surface properties, 177
surfactant, 164, 174
sustainability, 153, 155

T

Taiwan, 13, 85
target, ix, 4, 11, 16, 88, 81, 127, 157, 161
techniques, 3, 11, 12, 13, 45, 173
technologies, 1, 11, 44, 167, 181
technology, xiii, 3, 4, 11, 168, 169, 171, 173
temperature, 33, 36, 41, 42, 43, 82, 84, 87, 83, 160, 183
testing, 42
textural character, 83
Thailand, vii, xi, 1, 2, 3, 4, 5, 6, 8, 11, 12, 26, 27, 28, 81, 82, 83, 85, 93, 82, 84, 97, 101, 136, 138, 155, 172
time use, 88
titanium, 32, 169
toxic effect, 3
toxic metals, 12, 184
toxicity, 81, 153
transition metal, 18
transport, 89
transportation, 8, 41, 42

treatment methods, 4, 5, 146
tuff, 164
Turkey, 12, 33

U

UK, 174, 180
UN, 155
United, xiii, xv
United Nations, xiii
United States, xv
uranium, 175
USA, 27, 165, 168
USGS, 182

V

vacuum, 16
variables, 92
variations, 43, 44, 101
vomiting, 3, 12

W

Washington, 168, 180
waste, vii, 3, 4, 5, 6, 7, 15, 27, 28, 30, 31,
    32, 33, 37, 40, 41, 45, 83, 149, 163, 169,
    170, 171, 173, 174, 175, 176, 177, 178,
    180, 181, 182
waste disposal, 4, 6, 27
water purification, 6, 27
water quality, 166, 173
wood, 170, 176
wool, 6, 27, 38, 90, 165
worldwide, vii, 1, 22, 27, 42

Z

zeolites, 164, 166, 167, 173, 175, 183
zinc, 17, 165, 166, 167, 171, 172, 173, 175,
    177, 178, 182